Introduction to Food Engineering

Introduction to Food Engineering

Molly Drake

SYRAWOOD
PUBLISHING HOUSE

New York

Published by Syrawood Publishing House,
750 Third Avenue, 9th Floor,
New York, NY 10017, USA
www.syrawoodpublishinghouse.com

Introduction to Food Engineering
Molly Drake

International Standard Book Number: 978-1-64740-044-6 (Hardback)

Cataloging-in-Publication Data

Introduction to food engineering / Molly Drake.
    p. cm.
Includes bibliographical references and index.
ISBN 978-1-64740-044-6
1. Food industry and trade. 2. Food engineers. 3. Food--Biotechnology.
4. Agricultural engineering. 5. Food. I. Drake, Molly.
TP370 .I58 2020
664--dc23

TABLE OF CONTENTS

This book has been written, keeping in view that students want more practical information. Thus, my aim has been to make it as comprehensive as possible for the readers. I would like to extend my thanks to my family and co-workers for their knowledge, support and encouragement all along.

Food engineering is an interdisciplinary field, which combines diverse aspects of microbiology, applied physical sciences, chemistry and engineering for food and related industries. It is concerned with the application of principles of agricultural engineering, mechanical engineering and chemical engineering. It is also responsible for the development of technology that is essential for the cost-effective production and commercialization of food products and services. It includes various activities such as food processing, packaging, instrumentation, ingredient manufacturing, etc. Food engineering also uses advanced monitoring and control systems to facilitate automation and flexible manufacturing of food. This textbook is compiled in such a manner, that it will provide in-depth knowledge about the theory and practice of food engineering. It studies, analyses and upholds the pillars of this discipline and its utmost significance in modern times. Those in search of information to further their knowledge will be greatly assisted by this book.

A brief description of the chapters is provided below for further understanding:

Chapter – What is Food Engineering?

Food engineering is a discipline which deals with the application of microbiology, applied physical sciences, chemistry and engineering for food and related industries. It encompasses a wide range of activities which include food processing, food machinery, packaging, ingredient manufacturing, instrumentation, etc. This is an introductory chapter which will introduce briefly all the significant aspects of food engineering.

Chapter – Food Preservation

Food preservation is the process of treating and handling food in such a way that it minimizes the possibility of foodborne illnesses. Some of the diverse methods of food preservation are food drying, pickling, fermentation, food irradiation, brining and smoking. This chapter discusses in detail these methods of food preservation.

Chapter – High Pressure Processing of Food

High pressure food processing is a method of preservation and sterilization of food under high pressure. Such processing has numerous effects on food characteristics, food microorganisms and food components. The topics elaborated in this chapter will help in gaining a better perspective about high pressure food processing as well as its effects.

Chapter – Food Freezing

Food freezing is one of the methods of food preservation which inhibits microbial growth by lowering the temperature. It turns residual moisture into ice, thus inhibiting the growth of most bacterial species. This chapter closely examines the key concepts of food freezing such as thermal properties of frozen food and industrial freeze drying to provide an extensive understanding of the subject.

Chapter – Food Safety and Packaging

Food safety is concerned with handling, preparation and storage of food in such a way that it prevents food contamination. Food packaging refers to the enclosure of food to protect it from damage, spoilage and pest attacks. This chapter closely examines the key concepts of food safety and packaging as well as the materials which are commonly used to pack food.

Chapter – Processing of Common Food

The conversion of one form of food into another form is known as food processing. Some of the common foods which undergo processing before consumption are cereals, meats and milk. The topics elaborated in this chapter will help in gaining a better perspective about processing of these foods.

Molly Drake

CHAPTER 1

What is Food Engineering?

Food engineering is a discipline which deals with the application of microbiology, applied physical sciences, chemistry and engineering for food and related industries. It encompasses a wide range of activities which include food processing, food machinery, packaging, ingredient manufacturing, instrumentation, etc. This is an introductory chapter which will introduce briefly all the significant aspects of food engineering.

Food Engineering is a specialized sub field within Agricultural Engineering and is focused on the application of engineering to the production and distribution of food. In today's world, food production is a significant issue and Food Engineers play an important role in addressing that.

Food that is grown or processed must follow strict safety health and standards to ensure they are safe and that people do not get sick (or worse) from consuming produced food. Food Engineers control the health and safety of food production by designing and operating food processing plants. They also involve themselves in waste management and the genetic modification of foods. Food Engineers generally work for food manufacturing companies, but they can also work in the pharmaceutical and health care industries.

Genetically Modified Food

GM foods are derived from genetically modified organisms (GMOs), specifically plants and animals of agricultural importance. GMOs are defined as organisms whose genomes have been altered in ways that do not occur naturally. Although the definition of GMOs includes organisms that have been genetically modified by selective breeding, the most commonly used definition refers to organisms modified through genetic engineering or recombinant DNA technologies. Genetic engineering allows one or more genes to be cloned and transferred from one organism to another—either between individuals of the same species or between those of unrelated species. It also allows an organism's endogenous genes to be altered in ways that lead to enhanced or reduced expression levels. When genes are transferred between unrelated species, the resulting organism is called transgenic. The term cisgenic is sometimes used to describe gene transfers within a species. In contrast, the term biotechnology is a more general one, encompassing a wide range of methods that manipulate organisms or their components—such as isolating enzymes or producing wine, cheese, or yogurt. Genetic modification of plants or animals is one aspect of biotechnology.

In 2012, it was estimated that GM crops were grown in approximately 30 countries on 11 percent of the arable land on Earth. The majority of these GM crops (almost 90 percent) are grown in five countries—the United States, Brazil, Argentina, Canada, and India. Of these five, the United States accounts for approximately half of the acreage devoted to GM crops. According to the U.S. Department of Agriculture, 93 percent of soybeans and 88 percent of maize grown in the United States are from GM crops. In the United States, more than 70 percent of processed foods contain ingredients derived from GM crops.

Soon after the release of the Flavr Savr tomato in the 1990s, agribusinesses devoted less energy to designing GM foods to appeal directly to consumers. Instead, the market shifted toward farmers, to provide crops that increased productivity. By 2012, approximately 200 different GM crop varieties were approved for use as food or livestock feed in the United States. However, only about two dozen are widely planted. These include varieties of soybeans, corn, sugar beets, cotton, canola, papaya, and squash. Table lists some of the common GM food crops available for planting in the United States. Of these GM crops, by far the most widely planted are varieties that are herbicide tolerant or insect resistant.

Table: Some GM crops approved for food, feed, or cultivation.

Crop	Number of Varieties	GM Characteristics
Soybeans	19	Tolerance to glyphosate herbicide
		Tolerance to glufosinate herbicide
		Reduced saturated fats
		Reduced saturated fats
		Enhanced omega-3 fatty acid
Maize	68	Tolerance to glyphosate herbicide
		Tolerance to glufosinate herbicide
		Bt insect resistance
		Enhanced ethanol production
Cotton	30	Tolerance to glyphosate herbicide
		Bt insect resistance
Potatoes	28	Bt insect resistance
Canola	23	Tolerance to glyphosate herbicide
		Tolerance to glufosinate herbicide
		Enhanced lauric acid
Papaya	4	Resistance to papaya ringspot virus
Sugar beets	3	Tolerance to glyphosate herbicide
Rice	3	Tolerance to glufosinate herbicide
Zucchini squash	2	Resistance to zucchini, watermelon, and cucumber mosaic viruses
Alfalfa	2	Tolerance to glyphosate herbicide
Plum	1	Resistance to plum pox virus

Herbicide-resistant GM Crops

Weed infestations destroy about 10 percent of crops worldwide. To combat weeds, farmers often apply herbicides before seeding a crop and between rows after the crops are growing. As the most efficient broad-spectrum herbicides also kill crop plants, herbicide use may be difficult and limited. Farmers also use tillage to control weeds; however, tillage damages soil structure and increases erosion.

Herbicide-tolerant varieties are the most widely planted of GM crops, making up approximately 70 percent of all GM crops. The majority of these varieties contain a bacterial gene that confers tolerance to the broad-spectrum herbicide glyphosate—the active ingredient in commercial herbicides such as Roundup. Studies have shown that glyphosate is effective at low concentrations, is degraded rapidly in soil and water, and is not toxic to humans.

Farmers who plant glyphosate-tolerant crops can treat their fields with glyphosate, even while the GM crop is growing. This approach is more efficient and economical than mechanical weeding and reduces soil damage caused by repeated tillage. It is suggested that there is less environmental impact when using glyphosate, compared with having to apply higher levels of other, more toxic, herbicides.

Recently, evidence suggests that some weeds may be developing resistance to glyphosate, thereby reducing the effectiveness of glyphosate-tolerant crops.

Insect-resistant GM Crops

The second most prevalent GM modifications are those that make plants resistant to agricultural pests. Insect damage is one of the most serious threats to worldwide food production. Farmers combat insect pests using crop rotation and predatory organisms, as well as applying insecticides.

The most widely used GM insect-resistant crops are the Bt crops. Bacillus thuringiensis (Bt) is group of soildwelling bacterial strains that produce crystal (Cry) proteins that are toxic to certain species of insects. These Cry proteins are encoded by the bacterial cry genes and form crystal structures during sporulation. The Cry proteins are toxic to Lepidoptera (moths and butterflies), Diptera (mosquitoes and flies), Coleoptera (beetles), and Hymenoptera (wasps and ants). Insects must ingest the bacterial spores or Cry proteins in order for the toxins to act. Within the high pH of the insect gut, the crystals dissolve and are cleaved by insect protease enzymes. The Cry proteins bind to receptors on the gut wall, leading to breakdown of the gut membranes and death of the insect.

Each insect species has specific types of gut receptors that will match only a few types of Bt Cry toxins. As there are more than 200 different Cry proteins, it is possible to select a Bt strain that will be specific to one pest type.

Bt spores have been used for decades as insecticides in both conventional and organic gardening, usually applied in liquid sprays. Sunlight and soil rapidly break down the Bt insecticides, which have not shown any adverse effects on groundwater, mammals, fish, or birds. Toxicity tests on humans and animals have shown that Bt causes few negative effects.

The GM crop plants will then manufacture their own Bt Cry proteins, which will kill the target pest species when it eats the plant tissues.

Although Bt crops have been successful in reducing crop damage, increasing yields, and reducing the amounts of insecticidal sprays used in agriculture, they are also controversial. Early studies suggested that Bt crops harmed Monarch butterfly populations, although more recent studies have drawn opposite conclusions.

GM Crops for Direct Consumption

To date, most GM crops have been designed to help farmers increase yields. Also, most GM food crops are not consumed directly by humans, but are used as animal feed or as sources of processed food ingredients such as oils, starches, syrups, and sugars. For example, 98 percent of the U.S. soybean crop is used as livestock feed. The remainder is processed into a variety of food ingredients, such as lecithin, textured soy proteins, soybean oil, and soy flours. However, a few GM foods have been developed for direct consumption. Examples are rice, squash, and papaya.

One of the most famous and controversial examples of GM foods is Golden Rice—a rice variety designed to synthesize beta-carotene (the precursor to vitamin A) in the rice grain endosperm.

Vitamin A deficiency is a serious health problem in more than 60 countries, particularly countries in Asia and Africa. The World Health Organization estimates that 190 million of the world's children and 19 million pregnant women are vitamin A deficient. Between 250,000 and 500,000 children with vitamin A deficiencies become blind each year, and half of these will die within a year of losing their sight. As vitamin A is also necessary for immune system function, deficiencies lead to increases in many other conditions, including diarrhea and virus infections. The most seriously affected people live in the poorest countries and have a basic starch-centered diet, often mainly rice. Vitamin A is normally found in dairy products and can be synthesized in the body from beta-carotene found in orange-colored fruits and vegetables and in green leafy vegetables.

Several approaches are being taken to alleviate the vitamin A deficiency status of people in developing countries. These include supplying high-dose vitamin A supplements and growing fresh fruits and vegetables in home gardens. These initiatives have had partial success, but the expense of delivering education and supplementation has impeded the effectiveness of these programs.

In the 1990s, scientists began to apply recombinant DNA technology to help solve vitamin A deficiencies in people with rice-based diets. Although the rice plant naturally produces beta-carotene in its leaves, it does not produce it in the rice grain endosperm, which is the edible part of the rice. The beta-carotene precursor, geranylgeranyldiphosphate, is present in the endosperm, but the enzymes that convert it to beta-carotene are not synthesized.

In the first version of Golden Rice, scientists introduced the genes phytoene synthase (psy) cloned from the daffodil plant and carotene desaturase (crtI) cloned from the bacterium Erwinia uredovora into rice plants. The bacterial crtI gene was chosen because the enzyme encoded by this gene can perform the functions of two of the missing rice enzymes, thereby simplifying the transformation process. The resulting plant produced rice grains that were a yellow color due to the presence of beta-carotene. This strain synthesized modest levels of beta-carotene—but only enough to potentially supply 15–20 percent of the recommended daily allowance of vitamin A. In the second version of the GM plant, called Golden Rice 2, the daffodil psy gene was replaced with the psy gene from maize. Golden Rice 2 produced beta-carotene levels that were more than 20-fold greater than those in Golden Rice.

Clinical trials have shown that the beta-carotene in Golden Rice 2 is efficiently converted into vitamin A in humans and that about 150 grams of uncooked Golden Rice 2 (which is close to the normal daily rice consumption of children aged 4–8 years) would supply all of the childhood daily requirement for vitamin A.

At the present time, Golden Rice 2 is undergoing field, biosafety, and efficacy testing in preparation for approval by government regulators in Bangladesh and the Philippines. If Golden Rice 2 proves useful in alleviating vitamin A deficiencies and is approved for use, seed will be made available at the same price as non-GM seed and farmers will be allowed to keep and replant seed from their own crops.

Despite the promise of Golden Rice 2, controversies remain. Critics of GM foods suggest that Golden Rice could make farmers too dependent on one type of food or might have long-term health or environmental effects.

Methods used to Create GM Plants

Most GM plants are created using one of two approaches: the biolistic method or Agrobacterium tumefaciensmediated transformation technology. Both methods target plant cells that are growing in vitro. Scientists can generate plant tissue cultures from various types of plant tissues, and these cultured cells will grow in either liquid cultures or on the surface of solid growth media. When grown in the presence of specific nutrients and hormones, these cultured cells will form clumps of cells called calluses, which, when transferred to other types of media, will form roots. When the rooted plantlets are mature, they are transferred to soil medium in greenhouses where they develop into normal plants.

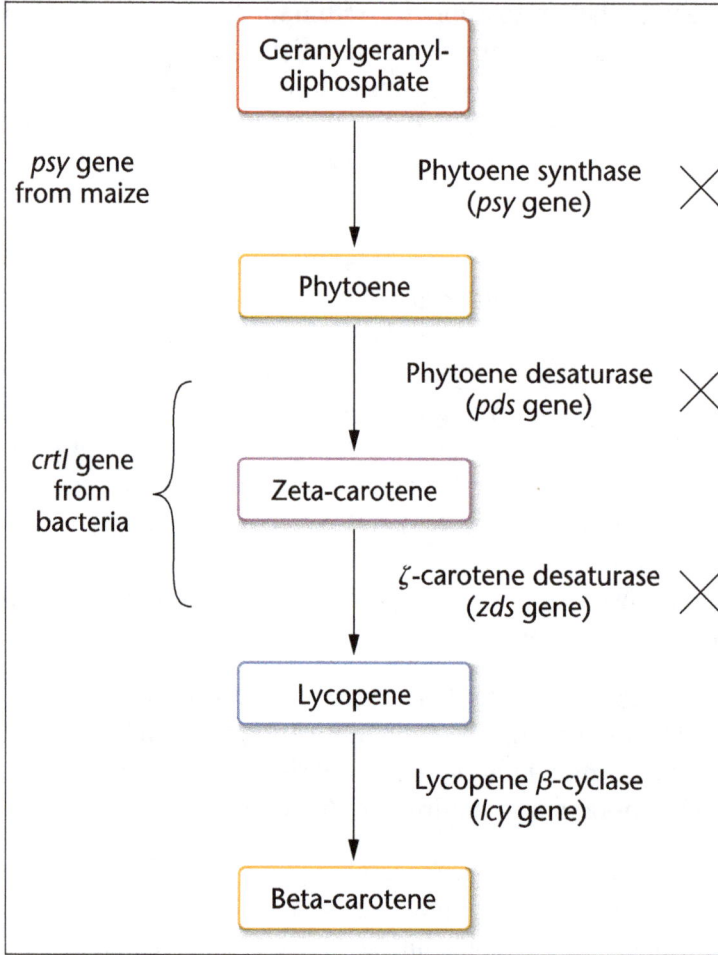

Beta-carotene pathway in Golden Rice 2. Rice plant enzymes and genes involved in beta-carotene synthesis are shown on the right. The enzymes that are not expressed in rice endosperm are indicated with an "X." The genes inserted into Golden Rice 2 are shown on the left.

The biolistic method is a physical method of introducing DNA into cells. Particles of heavy metals such as gold are coated with the DNA that will transform the cells; these particles are then fired at high speed into plant cells in vitro, using a device called a gene gun. Cells that survive the bombardment may take up the DNA-coated particles, and the DNA may migrate into the cell nucleus and integrate into a plant chromosome. Plants that grow from the bombarded cells are then selected for the desired phenotype.

Although biolistic methods are successful for a wide range of plant types, a much improved transformation rate is achieved using Agrobacterium-mediated technology. Agrobacterium tumefaciens (also called Rhizobium radiobacter) is a soil microbe that can infect plant cells and cause tumors.

Non-GM and Golden Rice 2. Golden Rice 2 contains high levels of
beta-carotene, giving the rice endosperm a yellow color. The intensity
of the color reflects the amount of beta-carotene in the endosperm.

These characteristics are conferred by a 200-kb tumor inducing plasmid called a Ti plasmid. After infection with Agrobacterium, the Ti plasmid integrates a segment of its DNA known as transfer DNA (T-DNA) into random locations within the plant genome. To use the Ti plasmid as a transformation vector, scientists remove the T-DNA segment and replace it with cloned DNA of the genes to be introduced into the plant cells.

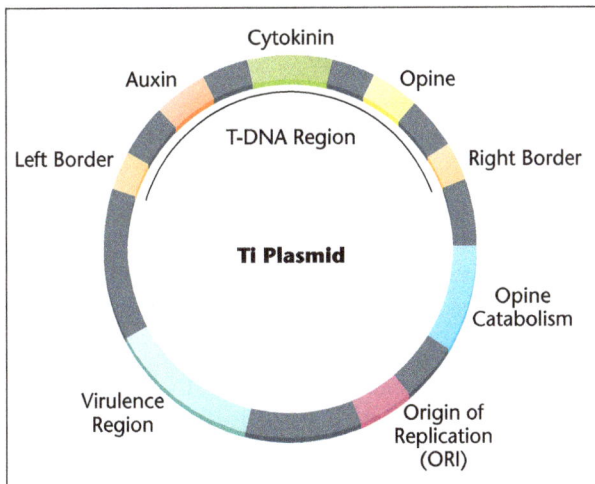

In order to have the newly introduced gene expressed in the plant, the gene must be cloned next to an appropriate promoter sequence that will direct transcription in the required plant tissue. For example, the beta-carotene pathway genes introduced into Golden Rice were cloned next to a promoter that directs transcription of the genes in the rice endosperm. In addition, the transformed gene requires appropriate transcription termination signals and signal sequences that allow insertion of the encoded protein into the correct cell compartment.

Structure of the Ti plasmid. The 250-kb Ti plasmid from Agrobacterium tumefaciens inserts the T-DNA portion of the plasmid into the host cell's nuclear genome and induces tumors. Genes within the virulence region code for enzymes responsible for transfer of T-DNA into the plant genome. The T-DNA region contains auxin and cytokinin genes that encode hormones responsible for cell growth and tumor formation. The opine genes encode compounds used as energy sources for the bacterium. The T-DNA region of the Ti plasmid is replaced with the gene of interest when the plasmid is used as a transformation vector.

Selectable Markers

The rates at which T-DNA successfully integrates into the plant genome and becomes appropriately expressed are low. Often, only one cell in 1000 or more will be successfully transformed. Before growing cultured plant cells into mature plants to test their phenotypes, it is important to eliminate the background of nontransformed cells. This can be done using either positive or negative selection techniques.

An example of negative selection involves use of a marker gene such as the hygromycin-resistance gene. This gene, together with an appropriate promoter, can be introduced into plant cells along with the gene of interest. The cells are then incubated in culture medium containing hygromycin—an antibiotic that also inhibits the growth of eukaryotic cells. Only cells that express the hygromycinresistance gene will survive. It is then necessary to verify that the resistant cells also express the cotransformed gene. This is often done by techniques such as PCR amplification using gene-specific primers. Plants that express the gene of interest are then tested for other characteristics, including the phenotype conferred by the introduced gene of interest.

An example of positive selection involves the use of a selectable marker gene such as that encoding phosphomannose isomerase (PMI). This enzyme is common in animals but is not found in most plants. It catalyzes the interconversion of mannose 6-phosphate and fructose 6-phosphate. Plant cells that express the pmi gene can survive on synthetic culture medium that contains only mannose as a carbon source. Cells that are cotransformed with the pmi gene under control of an appropriate promoter and the gene of interest can be positively selected by growing the plant cells on a mannose-containing medium. This type of positive selection was used to create Golden Rice 2. Studies have shown that purified PMI protein is easily digested, nonallergenic, and nontoxic in mouse oral toxicity tests. A variation in positive selection involves use of a marker gene whose expression results in a visible phenotype, such as deposition of a colored pigment.

The following descriptions illustrate the methods used to engineer two GM crops: Roundup-Ready soybeans and Golden Rice 2.

Roundup-ready Soybeans

The Roundup-ready soybean GM variety received market approval in the United States in 1996. It is a GM plant with resistance to the herbicide glyphosate, the active ingredient in Roundup, a commercially available broadspectrum herbicide. Glyphosate interferes with the enzyme 5-enolpyruvylshikimate-3-phosphate synthase (EPSPS), which is present in all plants and is necessary for plant synthesis of the aromatic amino acids phenylalanine, tyrosine, and tryptophan. EPSPS is not present in mammals, which obtain aromatic amino acids from their diets.

To produce a glyphosate-resistant soybean plant, researchers cloned an epsps gene from the Agrobacterium strain CP4. This gene encodes an EPSPS enzyme that is resistant to glyphosate. They then cloned the CP4 epsps gene downstream of a constitutively expressed promoter from the cytomegalovirus to allow gene expression in all plant tissues. In addition, a short peptide known as a chloroplast transit peptide (in this case from petunias) was cloned onto the 5' end of the epsps gene-coding sequence. This allowed newly synthesized EPSPS protein to be inserted into the soybean chloroplast. The final plasmid contained two CP4 epsps genes and, for the initial experiments, a beta-glucuronidase (GUS) gene from E. coli. The GUS gene acted as a positive marker, as cells that expressed the plasmid after transformation could be detected by the presence of a blue precipitate. The final cell line chosen for production of Roundup-Ready soybeans did not contain the GUS gene.

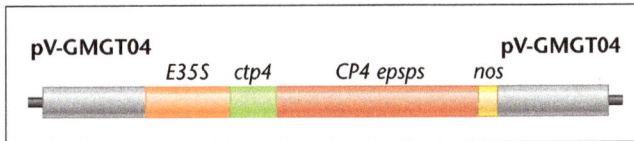

Portion of plasmid pV-GMGT04 used to create Roundup-Ready soybeans. A 1365-bp fragment encoding the EPSPS enzyme from Agrobacterium CP4 was cloned downstream from the cauliflower mosaic virus E35S promoter and the petunia chloroplast transit peptide signal sequence (ctp4). CTP4 signal sequences direct the EPSPS protein into chloroplasts, where aromatic amino acids are synthesized. The CP4 epsps coding region was cloned upstream of the nopaline synthase (nos) transcription termination and polyadenylation sequences. The CP4 epsps sequences encode a 455-amino-acid 46-kDa ESPSP protein.

The plasmids were introduced into cultured soybean cells using biolistic bombardment. Afterward, cells were treated with glyphosate to eliminate any nontransformed cells. The resulting calluses were grown into plants, which were then field tested for glyphosate resistance and a large number of other parameters, including composition, toxicity, and allergenicity.

Golden Rice 2

To create Golden Rice 2, scientists cloned three genes into the T-DNA region of a Ti plasmid. The Ti plasmid, called pSYN12424, is shown in figure. The first gene was the

carotene desaturase (crtI) gene from Erwinia uredovora, fused between the rice glutelin gene promoter (Glu) and the nos gene terminator region (nos). The Glu promoter
directs transcription of the fusion gene specifically in the rice endosperm. The nos terminator was cloned from the Agrobacterium tumefaciens nopaline synthase gene and
supplies the transcription termination and polyadenylation sequences required at the
3' end of plant genes. The second gene was the phytoene synthase (psy) gene cloned
from maize. The maize psy gene has approximately 90 percent sequence similarity to
the rice psy gene and is involved in carotenoid synthesis in maize endosperm. This gene
was also fused to the Glu promoter and the nos terminator sequences in order to obtain
proper transcription initiation and termination in rice endosperm. The third gene was
the selectable marker gene, phosphomannose isomerase (pmi), cloned from E. coli. In
the Golden Rice 2 Ti plasmid, the pmi gene was fused to the maize polyubiquitin gene
promoter (Ubi1) and the nos terminator sequences. The Ubi1 promoter is a constitutive
promoter, directing transcription of the pmi gene in all plant tissues.

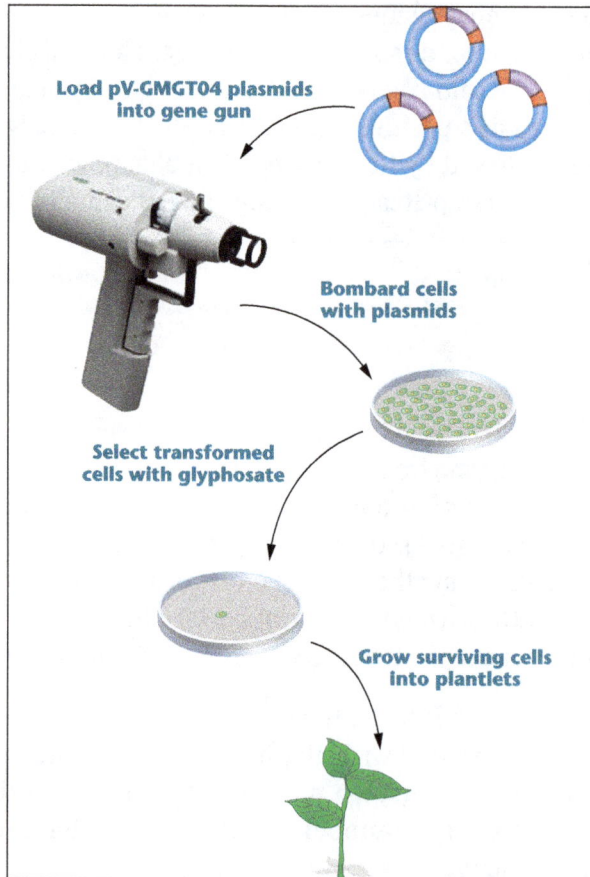

Method for creating Roundup-Ready soybeans. Plasmids were loaded into the gene
gun and fired at high pressure into cells growing in tissue cultures. Cells were grown in
the presence of glyphosate to select those that had integrated and expressed the epsps
gene. Surviving cells were stimulated to form calluses and to grow into plantlets.

| Glu | **crtI** | nos | Glu | **psy** | nos | Ubi1 | **pmi** | nos |

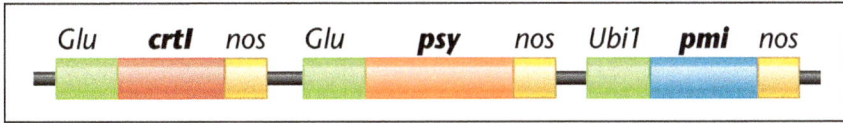

T-DNA region of T1 plasmid pSYN12424. The Ti plasmid used to create Golden Rice 2 contained the carotene desaturase (crtI) gene cloned from bacteria, the phytoene synthase (psy) gene cloned from maize, and the phosphomannose isomerase (pmi) gene cloned from E. coli. The glutelin (Glu) gene promoter directs transcription in rice endosperm, and the polyubiquitin (Ubi1) promoter directs transcription in all tissues. Transcription termination signals were provided by the nopaline synthase (nos) gene 3' region.

To introduce the pSYN12424 plasmid into rice cells, researchers established embryonic rice cell cultures and infected them with Agrobacterium tumefaciens that contained pSYN12424. The cells were then placed under selection, using culture medium containing only mannose as a carbon source. Surviving cells expressing the pmi gene were then stimulated to form calluses that were grown into plants. To confirm that all three genes were present in the transformed rice plants, samples were taken and analyzed by the polymerase chain reaction (PCR) using gene-specific primers. Plants that contained one integrated copy of the transgenic construct and synthesized betacarotene in their seeds were selected for further testing.

GM Foods Controversies

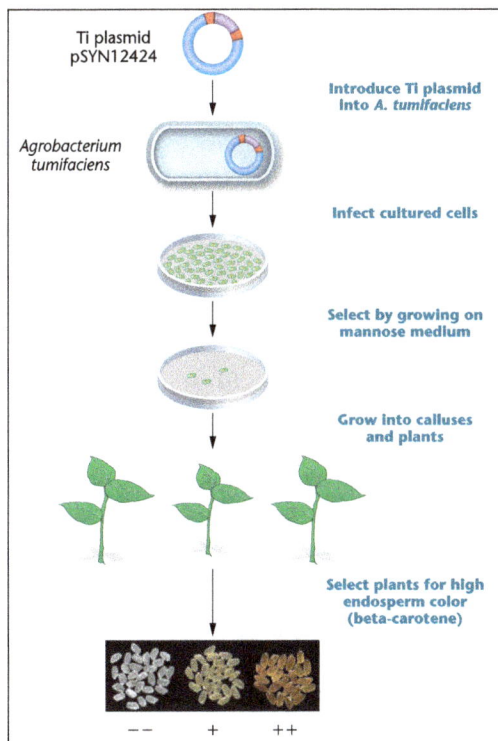

Ti plasmid pSYN12424

Introduce Ti plasmid into *A. tumifaciens*

Agrobacterium tumifaciens

Infect cultured cells

Select by growing on mannose medium

Grow into calluses and plants

Select plants for high endosperm color (beta-carotene)

−−   +   ++

Method for creating Golden Rice 2: Rice plant cells were transformed by pSYN12424 and selected on mannose-containing medium. Plants that produced high levels of beta-carotene in rice grain endosperm (+ +), based on the intensity of the grain's yellow color, were selected for further analysis.

GM foods may be the most contentious of all products of modern biotechnology. Advocates of GM foods state that the technologies have increased farm productivity, reduced pesticide use, preserved soils, and have the potential to feed growing human populations. Critics claim that GM foods are unsafe for both humans and the environment; accordingly, they are applying pressure on regulatory agencies to ban or severely limit the extent of GM food use. These campaigns have affected regulators and politicians, resulting in a patchwork of regulations throughout the world. Often the debates surrounding GM foods are highly polarized and emotional, with both sides in the debate exaggerating their points of view and selectively presenting the data.

One point that is important to make as we try to answer this question is that it is not possible to make general statements about all "GM foods." Each GM crop or organism contains different genes from different sources, attached to different expression sequences, accompanied by different marker or selection genes, inserted into the genome in different ways and in different locations. GM foods are created for different purposes and are used in ways that are both planned and unplanned. Each construction is unique and therefore needs to be assessed separately. We will now examine two of the main GM foods controversies: those involving human health and safety, and environmental effects.

Health and Safety

GM food advocates often state that there is no evidence that GM foods currently on the market have any adverse health effects, either from the presence of toxins or from potential allergens. These conclusions are based on two observations. First, humans have consumed several types of GM foods for more than 20 years now, and no reliable reports of adverse effects have emerged. Second, the vast majority of toxicity tests in animals, which are required by government regulators prior to approval, have shown no negative effects. A few negative studies have been published, but these have been criticized as poorly executed or nonreproducible.

Critics of GM foods counter the first observation in several ways. First, as described previously, few GM foods are eaten directly by consumers. Instead, most are used as livestock feed, and the remainder form the basis of purified food ingredients. Although no adverse effects of GM foods in livestock have been detected, the processing of many food ingredients removes most, if not all, plant proteins and DNA. Hence, ingestion of GM food-derived ingredients may not be a sufficient test for health and safety. Second, GM foods critics argue that there have been few human clinical trials to directly examine the health effects of most GM foods. One notable exception is Golden Rice 2, which has undergone two small clinical trials. They also say that the toxicity studies that have been completed are performed in animals—primarily rats and mice—and most of these are short-term toxicity studies.

Supporters of GM foods answer these criticisms with several other arguments. The first argument is that shortterm toxicity studies in animals are well-established methods for detecting toxins and allergens. The regulatory processes required prior to approval of any GM food demand data from animal toxicity studies. If any negative effects are detected, approval is not given. Supporters also note that several dozen long-term toxicity studies have been published that deal with GM crops such as glyphosate-resistant soybeans and Bt corn, and none of these has shown longterm negative effects on test animals. A few studies that report negative long-term effects have been criticized as poorly designed and unreliable. GM food advocates note that human clinical trials are not required for any other food derived from other genetic modification methods such as selective breeding. During standard breeding of plants and animals, genomes may be mutagenized with radiation or chemicals to enhance the possibilities of obtaining a desired phenotype. This type of manipulation has the potential to introduce mutations into genes other than the ones that are directly selected. Also, plants and animals naturally exchange and shuffle DNA in ways that cannot be anticipated. These include interspecies DNA transfers, transposon integrations, and chromosome modifications. These events may result in unintended changes to the physiology of organisms—changes that could potentially be as great as those arising in GM foods.

Environmental Effects

Critics of GM foods point out that GMOs that are released into the environment have both documented and potential consequences for the environment—and hence may indirectly affect human health and safety. GM food advocates argue that these potential environmental consequences can be identified and managed. Here, we will describe two different aspects of GM foods as they may affect the natural environment and agriculture.

Emerging herbicide and insecticide resistance. Many published studies report that the planting of herbicidetolerant and insect-resistant GM crops has reduced the quantities of herbicides and insecticides that are broadly applied to agricultural crops. As a result, the effects of GM crops on the environment have been assumed to be positive. However, these positive effects may be transient, as herbicide and insecticide resistance is beginning to emerge.

Since glyphosate-tolerant crops were introduced in the mid-1990s, more than 24 glyphosate-resistant weed species have appeared in the United States. Resistant weeds have been found in 18 other countries, and in some cases, the presence of these weeds is affecting crop yields. One reason for the rapid rise of resistant weeds is that farmers have abandoned other weed-management practices in favor of using a single broad-spectrum herbicide. This strong selection pressure has brought the rapid evolution of weed species bearing gene variants that confer herbicide resistance. In response, biotechnology companies are developing new GM crops with tolerance to multiple herbicides. However, scientists argue that weeds will also develop resistance to the use of multiple herbicides, unless farmers vary their weed management practices and incorporate tillage, rotation, and other herbicides along with using the GM crop. Scientists point out that herbicide resistance is not limited to the use of GM crops. Weed populations will evolve resistance to any herbicide used to control them, and the speed of evolution will be affected by the extent to which the herbicide is used.

Herbicide-resistant weeds. Water hemp weeds, resistant to glyphosate herbicide, growing in a field of Roundup-Ready soybeans.

Since 1996, more than eight different species of insect pests have evolved some level of resistance to Bt insecticidal proteins. For example, in 2011 scientists reported the first cases of resistance of the western corn rootworm to Bt maize expressing the cry3Bb1 gene, in maize fields in Iowa. In 2010, scientists from Monsanto detected large numbers of pink bollworms with resistance to the toxin expressed from the cry1Ac gene in one variety of Bt cotton. In order to slow down the development of Bt resistance, several strategies are being followed. The first is to develop varieties of GM crops that express two Bt toxins simultaneously. Several of these varieties are already on the market and are replacing varieties that express only one Bt cry gene. The second strategy involves the use of "refuges" surrounding fields that grow Bt crops. These refuges contain non-GM crops. Insect pests grow easily within the refuges, which place no evolutionary pressure on the insects for resistance to Bt toxins. The idea is for these nonselected insects to mate with any resistant insects that appear in the Bt crop region of the field. The resulting hybrid offspring will be heterozygous for any resistance gene variant. As long as the resistance gene variant is recessive, the hybrids will be killed by eating the Bt crop. In fields that use refuges and plant GM crops containing two Bt genes, resistance to Bt toxins has been delayed or is absent. As with emerging herbicide resistance, farmers are also encouraged to combine the use of Bt crops with conventional pest control methods.

The spread of GM crops into non-GM crops. There have been several documented cases of GM crop plants appearing in uncultivated areas in the United States, Canada, Australia, Japan, and Europe. For example, GM sugar beet plants have been found growing in commercial top soils. GM canola plants have been found growing in ditches and along roadways, railway tracks, and in fill soils, far from the fields in which they were grown. A 2011 study found "feral" GM canola plants growing in 288 of 634 sample sites along roadways in North Dakota. Of these plants, 41 percent contained the CP4 EPSPS protein (conferring glyphosate resistance), and 39 percent contained the PAT protein (conferring resistance to the herbicide glufosinate). In addition, two of the plants (0.7 percent of the sample) expressed both proteins (resistant to both herbicides). GM plants that express both proteins have not been created by genetic modification and were assumed to have arisen by cross-fertilization of the other two GM crops. The researchers who conducted this survey were not surprised to find GM canola along transportation routes, as seeds are often spilled during shipping. More surprising was the extent of the distribution and the presence of hybridized GM canola plants.

One of the major concerns about the escape of GM crop plants from cultivation is the possibility of outcrossing or gene flow—the transfer of transgenes from GM crops into sexually compatible non-GM crops or wild plants, conferring undesired phenotypes to the other plants. Gene flow between GM crops and adjacent non-GM crops is of partic-ular concern for farmers who want to market their crops as "GM-free" or "organic" and for farmers who grow seed for planting.

Gene flow of GM transgenes has been documented in GM and non-GM canola as well as sugar beets, and in experiments using rice, wheat, and maize. GM critics often refer to controversial studies about GM outcrossing in Oaxaca, Mexico. In the first study in 2001, it was reported that the local maize crops contained transgenes from Monsanto's Roundup-Ready and Bt insect-resistant maize. As GM crops were not approved for use in Mexico, it was thought that the transgenes came from maize that had been imported from the United States as a foodstuff, and then had been planted by farmers who were not aware that the seeds were transgenic. Over the next ten years, subsequent studies reported mixed results. In some studies, the transgenes were not detected, and in others, the same transgenes were detected. There is still no consensus about whether gene flow has occurred between the GM and non-GM maize in Mexico.

It is thought that the presence of glyphosateresistant transgenes in wild plant popu-lations is not likely to be an environmental risk and would confer no positive fitness benefits to the hybrids. The presence of glyphosate-resistant genes in wild populations would, however, make it more difficult to eradicate the plants. This is illustrated in a case of escaped GM bentgrass in Oregon, where it has been difficult to get rid of the plants because it is no longer possible to use the relatively safe herbicide glyphosate. The potential for environmental damage may be greater if the GM transgenes did con-fer an advantage—such as insect resistance or tolerance to drought or flooding.

In an attempt to limit the spread of transgenes from GM crops to non-GM crops, regulators are considering a requirement to separate the crops so that pollen would be less likely to travel between them. Each crop plant would require different isolation distances to take into account the dynamics of pollen spreading. Several other methods are being considered. For example, one proposal is to make all GM plants sterile using RNAi technology. Another is to introduce the transgenes into chloroplasts. As chloroplasts are inherited maternally, their genomes would not be transferred via pollen. All of these containment methods are in development stages and may take years to reach the market.

Pros

Genetic modification can make crops more resistant to diseases as they grow.

Manufacturers use genetic modification to give foods desirable traits. For example, they have designed two new varieties of apple that turn less brown when cut or bruised.

The reasoning usually involves making crops more resistant to diseases as they grow. Manufacturers also engineer produce to be more nutritious or tolerant of herbicides.

Crop protection is the main rationale behind this type of genetic modification. Plants that are more resistant to diseases spread by insects or viruses result in higher yields for farmers and a more attractive product. Genetically modification can also increase nutritional value or enhance flavor.

All of these factors contribute to lower costs for the consumer. They can also ensure that more people have access to quality food.

Cons

Because genetically engineering foods is a relatively new practice, little is known about the long-term effects and safety. There are many purported downsides, but the evidence varies, and the main health issues associated with GMO foods are hotly debated. Research is ongoing.

Allergic Reactions

Some people believe that GMO foods have more potential to trigger allergic reactions. This is because they may contain genes from an allergen — a food that prompts an allergic reaction.

The World Health Organization (WHO) discourage genetic engineers from using DNA from allergens unless they can prove that the gene itself does not cause the problem. It is worth noting that there have been no reports of allergic effects of any GMO foods currently on the market.

Cancer

Some researchers believe that eating GMO foods can contribute to the development of cancer. They argue that because the disease is caused by mutations in DNA, it is dangerous to introduce new genes into the body.

The American Cancer Society (ACS) have said that there is no evidence for this. However, they note that no evidence of harm is not the same as proof of safety and that reaching a conclusion will require more research.

Antibacterial Resistance

There is concern that genetic modification, which can boost a crop's resistance to disease or make it more tolerant to herbicides, could affect the ability of people to defend against illness.

There is a small chance that the genes in food can transfer to cells the body or bacteria in the gut. Some GMO plants contain genes that make them resistant to certain antibiotics. This resistance could pass on to humans.

There is growing concern globally that people are becoming increasingly resistant to antibiotics. There is a chance that GMO foods could be contributing to this crisis.

The WHO have said that the risk of gene transfer is low. As a precaution, however, it has set guidelines for the manufacturers of GMO foods.

Outcrossing

Outcrossing refers to the risk of genes from certain GMO plants mixing with those of conventional crops.

There have been reports of low levels of GMO crops approved as animal feed or for industrial use being found in food meant for human consumption.

FOOD TECHNOLOGY

Food Technology is a science which deals with the techniques and principles involved in processing and preserving the food substances. The application of food science helps in manufacturing safe, wholesome and nutritious food products. The study of food technology is to develop new methods and systems for keeping food products safe and resistant from natural harms such as bacteria and other micro-organisms. Food processing helps in preservation enhances the flavor and reduces the toxins in the food product which results in better distributional efficiency and easy marketing of the food product.

The modern food processing techniques is the key to flourishing supermarkets we have today. Extra nutrients can be added while processing the food and processed food is less susceptible to spoilage. Some of the techniques used are spray drying, juice concentrates, freeze drying and the introduction of artificial sweeteners, colorants, and preservatives. Of late, many products such as dried instant soups, reconstituted fruits and juices, and self cooking meals were developed for the convenience of working people.

The food processing industries are involved in processes such as primary and secondary processing, preservation, quality management, packaging and labeling of a variety of products such as dairy products, fish products, fruit & vegetable products, meat & poultry products, confectionery products and food grains.

Few specializations in this fields are:

- Dairy,
- Sugar,
- Alcohol,
- Bakery and confectionery items,
- Oil and oil seed processing,

- Fruits and vegetables,

- Meat,

- Cereals.

FOOD PROCESSING

Food processing is the set of methods and techniques used to transform raw ingredients into food or food into other forms for consumption by humans or animals either in the home or by the food processing industry. Food processing typically takes clean, harvested crops or slaughtered and butchered animal products and uses these to produce attractive, marketable, and often long-life food products. Similar processes are used to produce animal feed. Extreme examples of food processing include the expert removal of toxic portions of the fugu fish or preparing space food for consumption under zero gravity.

The benefits of food processing include the preservation, distribution, and marketing of food, protection from pathogenic microbes and toxic substances, year-round availability of many food items, and ease of preparation by the consumer. On the other hand, food processing can lower the nutritional value of foods, and processed foods may include additives (such as colorings, flavorings, and preservatives) that may have adverse health effects.

Food processing dates back to prehistoric ages, with crude processing methods that included slaughtering, fermenting, sun drying, preserving with salt, and various means of cooking (such as roasting, smoking, steaming, and oven baking). Salt-preservation was especially common for foods that constituted the diets of warriors and sailors, up until the introduction of canning methods. Evidence for the existence of these methods exists in the writings of the ancient Greek, Chaldean, Egyptian, and Roman civilizations, as well as archaeological evidence from Europe, North and South America, and Asia. These tried and tested processing techniques remained essentially the same until the advent of the Industrial Revolution. Examples of ready-meals also exist from the period before the Industrial Revolution, such as the Cornish pasty and Haggis.

Modern food processing technology was largely developed in the nineteenth and twentieth centuries, to serve military needs. In 1809, Nicolas Appert invented a vacuum bottling technique that would supply food for French troops, and this contributed to the development of tinning and then canning by Peter Durand in 1810. Although initially expensive and somewhat hazardous due to the lead used in cans, canned goods later became a staple around the world. Pasteurization, discovered by Louis Pasteur in 1862, was a significant advance in ensuring the microbiological safety of food.

In the twentieth century, World War II, the space race, and the rising consumer society in developed countries contributed to the growth of food processing with such advances as spray drying, juice concentrates, freeze drying, and the introduction of artificial sweeteners, coloring agents, and preservatives such as sodium benzoate. In the late twentieth century, products such as dried instant soups, reconstituted fruits and juices, and self-cooking meals (such as "Meal, Ready-to-Eat," or MRE, field rations) were developed.

In Western Europe and North America, the second half of the twentieth century witnessed a rise in the pursuit of convenience, as food processors marketed their products especially to middle-class working wives and mothers. Frozen foods found their success in the sales of juice concentrates and "TV dinners." Processors utilized the perceived value of time to appeal to the postwar population, and this same appeal contributes to the success of convenience foods today.

Food Processing Methods

Beer fermenting at a brewery.

Common food processing techniques include:

- Removal of unwanted outer layers, such as potato peeling or the skinning of peaches.
- Chopping or slicing, such as to produce diced carrots.
- Mincing and macerating.
- Liquefaction, such as to produce fruit juice.
- Fermentation, as in beer breweries.
- Emulsification.
- Cooking, by methods such as baking, boiling, broiling, frying, steaming, or grilling.
- Mixing.
- Addition of gas such as air entrainment for bread or gasification of soft drinks.

- Proofing.

- Spray drying.

- Pasteurization.

- Packaging.

Performance Parameters for Food Processing

When designing processes for the food industry, the following performance parameters may be taken into account:

- Hygiene: It is measured, for instance, by the number of microorganisms per ml of finished product.

- Energy consumption: It is measured, for instance, by "ton of steam per ton of sugar produced".

- Minimization of waste: It is measured, for instance, by the "percentage of peeling loss during the peeling of potatoes".

- Labor used: It is measured, for instance, by the "number of working hours per ton of finished product".

- Minimization of cleaning stops: It is measured, for instance, by the "number of hours between cleaning stops".

Benefits

Microwave oven.

More and more people live in the cities far away from where food is grown and produced. In many families, the adults are work from home and therefore there is little time for the preparation of food based on fresh ingredients. The food industry offers products that fulfill many different needs: from peeled potatoes that simply need to be boiled at home to fully prepared ready meals that can be heated up in the microwave oven in a few minutes.

Benefits of food processing include toxin removal, preservation, easing marketing and distribution tasks, and increasing food consistency. In addition, it increases seasonal availability of many foods, enables transportation of delicate perishable foods across long distances, and makes many kinds of foods safe to eat by de-activating spoilage and pathogenic micro-organisms. Modern supermarkets would not be feasible without modern food processing techniques, long voyages would not be possible, and military campaigns would be significantly more difficult and costly to execute.

Modern food processing also improves the quality of life for allergy sufferers, diabetics, and other people who cannot consume some common food elements. Food processing can also add extra nutrients such as vitamins.

Processed foods are often less susceptible to early spoilage than fresh foods, and are better suited for long distance transportation from the source to the consumer. Fresh materials, such as fresh produce and raw meats, are more likely to harbor pathogenic microorganisms (for example, Salmonella) capable of causing serious illnesses.

Drawbacks

In general, fresh food that has not been processed other than by washing and simple kitchen preparation, may be expected to contain a higher proportion of naturally occurring vitamins, fiber and minerals than the equivalent product processed by the food industry. Vitamin C, for example, is destroyed by heat and therefore canned fruits have a lower content of vitamin C than fresh ones.

Corn Syrup being moved by tank car.

Food processing can lower the nutritional value of foods. Processed foods tend to include food additives, such as flavorings and texture enhancing agents, which may have little or no nutritive value, and some may be unhealthy. Some preservatives added or created during processing, such as nitrites or sulfites, may cause adverse health effects.

Processed foods often have a higher ratio of calories to other essential nutrients than unprocessed foods, a phenomenon referred to as "empty calories." Most junk foods are processed, and fit this category.

High quality and hygiene standards must be maintained to ensure consumer safety, and failure to maintain adequate standards can have serious health consequences.

Processing food is a very costly process, thus increasing the prices of foods products.

Trends in Modern Food Processing

Health

- Reduction of fat content in final product, for example, by using baking instead of deep-frying in the production of potato chips.

- Maintaining the natural taste of the product, for example, by using less artificial sweetener.

Hygiene

The rigorous application of industry and government endorsed standards to minimize possible risk and hazards. In the U.S., the standard adopted is HACCP.

Efficiency

- Rising energy costs lead to increasing usage of energy-saving technologies, for example, frequency converters on electrical drives, heat insulation of factory buildings, and heated vessels, energy recovery systems.

- Factory automation systems (often Distributed control systems) reduce personnel costs and may lead to more stable production results.

Industries

Food processing industries and practices include the following:

- Cannery,
- Industrial rendering,
- Meat packing plant,
- Slaughterhouse,
- Sugar industry,
- Vegetable packing plant.

PULSED ELECTRIC FIELD PROCESSING

Pulsed electric fields (PEF) is a non-thermal method of food preservation that uses short pulses of electricity for microbial inactivation and causes minimal detrimental effect on food quality attributes. PEF technology aims to offer consumers high-quality foods. For food quality attributes, PEF technology is considered superior to traditional thermal processing methods because it avoids or greatly reduces detrimental changes in the sensory and physical properties of foods. PEF technology aims to offer consumers high-quality foods. For food quality attributes, PEF technology is considered superior to traditional thermal processing methods because it avoids or greatly reduces detrimental changes in the sensory and physical properties of foods.

PEF technology has been presented as advantageous in comparison to, for instance, heat treatments, because it kills microorganisms while better maintaining the original color, flavor, texture, and nutritional value of the unprocessed food. PEF technology involves the application of pulses of high voltage to liquid or semi-solid foods placed between two electrodes. Most PEF studies have focused on PEF treatments effects on the microbial inactivation in milk, milk products, egg products, juice and other liquid foods.

Pulsed electric field technology (PEF) is viewed as one of the most promising nonthermal methods for inactivating microorganisms in foods. Electric fields in the range of 5-50 kV/cm generated by the application of short high voltage pulses (μs) between two electrodes cause microbial inactivation at temperatures below those used in thermal processing. The precise mechanisms by which microorganisms are inactivated by pulsed electric fields are not well understood; however, it is generally accepted that PEF leads to the penneabilization of microbial membranes.

Non-thermal processes have gained importance in recent years due to the increasing demand for foods with a high nutritional value and fresh-like characteristics, representing an alternative to conventional thermal treatments. Pulsed electric fields (PEF) is an emerging technology that has been extensively studied for non-thermal food processing. PEF processing has been studied by a number of researchers across a wide range of liquid foods. Apple and orange juices are among the foods most often treated in PEF studies. The sensory attributes of juices are reported to be well preserved, and the shelf life is extended. Yogurt drinks, apple sauce, and salad dressing have also been shown to retain a fresh-like quality with extended shelf life after processing. Other PEF-processed foods include milk, tomato juice, carrot juice, pea soup, liquid whole egg, and liquid egg products.

Food preservation technologies are based on the prevention of microbial growth or on the microbial inactivation. In many cases, foods are preserved by inhibiting microbial activity through those factors that most effectively influence the growth and survival of microorganisms such as temperature, water activity, addition of preservatives, pH, and modified atmosphere. In this case, the microorganisms will not be destroyed and will still be metabolically active and viable if transferred to favorable conditions. As estimates of the infection dose of some pathogenic microorganisms are very low, growth of these microorganisms in foods is not necessary to cause infection.

To qualify as an alternative method, a new technology should have significant impact on quality while at the same time maintain the cost of technology within feasibility limits. In recent years, several technologies have been investigated that have the capability of inactivating microorganisms at lower temperatures than typically used in conventional heat treatments.

Application of pulsed electric fields of high intensity and duration from microseconds to milliseconds may cause temporary or permanent permeabilization of cell membranes. The effects of PEF on biomembranes have been thoroughly studied since the use of PEF has attracted great interest in several scientific areas such as cell biology, biotechnology, medicine, or food technology.

With those obtained for other thermal and non-thermal processing technologies, with a special stress on the effect of PEF-processing variables on the bioactive composition of foods throughout their whole shelf-life. Furthermore, different examples are presented to illustrate not only the potential but also the limitations of PEF technology when aiming at preserving the health-promoting features of plant-based foods. With the use of electric fields, PEF technology enables inactivation of vegetative cells of bacteria and yeasts in various foods. As bacterial spores are resistant to pulsed electric fields, applications of this technology mainly focus on food-borne pathogens and spoilage microorganisms, especially for acidic food products. In addition to the volumetric effect of PEF technology in controlling the microbiological safety of foods in a fast and homogenous manner, successful application provides extended shelf life without the use of heat to preserve the sensory and nutritional value of foods. PEF technology has the potential to economically and efficiently improve energy usage, besides the advantage of providing microbiologically safe and minimally processed foods. Successful application of PEF technology suggests an alternative substitute for conventional thermal processing of liquid food products such as fruit juices, milk, and liquid egg.

Nonthermal Technologies for Food Processing

Nonthermal technologies represent a novel area of food processing and are currently being explored on a global scale; research has grown rapidly in the last few years in particular. The main purpose of thermal processing is the inactivation of pathogenic microorganisms and spores (depending on the treatment) to provide consumers with a microbiologically safe product. However, despite the benefits of thermal treatment, a number of changes take place in the product that alter its final quality, for example, flavor, color, texture, and general appearance. Now, consumers are looking for fresh-like characteristics in their food, along with high sensorial quality and nutrient content. Consumers are more aware of food content and the technologies used to process their food, showing a higher preference for natural products free of chemicals and additives. Thus, the need for processing alternatives that can achieve microbial inactivation, preserve food's fresh like characteristics, and provide environment friendly products, all at a reasonable cost, has become the present challenge of numerous food scientists/technologists around the world.

Nonthermal processing technologies were designed to eliminate the use of elevated temperatures during processing and so avoid the adverse effects of heat on the flavour, appearance and nutritive value of foods.

Novel nonthermal processes, such as high hydrostatic pressure (HHP), pulsed electric fields (PEFs), ionizing radiation and ultrasonication, are able to inactivate microorganisms at ambient or sublethal temperatures. Many of these processes require very high treatment intensities, however, to achieve adequate microbial destruction in low-acid foods. Combining nonthermal processes with conventional preservation methods enhances their antimicrobial effect so that lower process intensities can be used. Combining two or more nonthermal processes can also enhance microbial inactivation and allow the use of lower individual treatment intensities. For conventional preservation treatments, optimal microbial control is achieved through the hurdle concept, with synergistic effects resulting from different components of the microbial cell being targeted simultaneously. The mechanisms of inactivation by nonthermal processes are still unclear; thus, the bases of synergistic combinations remain speculative.

Nonthermal technologies encompass all preservation treatments that are effective at ambient or sub lethal temperatures including antimicrobial additives, pH adjustment and modified atmospheres. The term 'nonthermal processing' is more apt for novel nonthermal technologies, such as high hydrostatic pressure, pulsed electric fields (PEFs), high-intensity ultrasound, ultraviolet light, pulsed light, ionizing radiation and oscillating magnetic fields, which are intended for application as microbe-inactivating processes during food manufacture. Such novel technologies have the ability to inactivate microorganisms to varying degrees.

One nonthermal technology, high hydrostatic pressure (HHP), has shown a negligible effect on the nutrient content of food, for example, in processing of fruits and vegetables, where pressure has minimal effect on the anthocyanin content after processing. Anthocyanins are considered phytonutrients, and they not only are responsible for color but also have an important antioxidant effect on human health. However, anthocyanin content in juices after pulsed electric fields (PEF) treatment has shown contradictory results. Some researchers report a minimum effect on the pigment content after processing, while others show that there is degradation in anthocyanin content after pulsing.

The most extensively researched and promising nonthermal processes appear to be high hydrostatic pressure (HHP), pulsed electric fields (PEF) and high intensity ultrasound combined with pressure. Gamma irradiation has high potential although its development and commercialization has been hampered in the past by unfavourable public perceptions.

Despite the current gaps in understanding, combining nonthermal processes with other nonthermal technologies has been investigated to improve control over food borne microorganisms, with promising results. A better understanding of the antimicrobial mechanisms of emerging nonthermal technologies as well as their effectiveness when combined with traditional food preservation hurdles is needed so that new food preservation strategies can be developed on a sound scientific basis.

High-pressure processing applied at room temperature yields a product with most of food's quality attributes intact; for example, pressurization does not affect covalent bonds, avoiding any development of strange flavors in the food.

Ultrasound has also been used in milk pasteurization, with important results; milk shows a higher degree of homogenization, whiter color, and better stability after processing. In this method, pasteurization and homogenization are completed in a one-step process.

Principles of Pulsed Electric Field

The basic principle of the PEF technology is the application of short pulses of high electric fields with duration of microseconds micro- to milliseconds and intensity in the order of 10-80 kV/cm. The processing time is calculated by multiplying the number of pulses times with effective pulse duration. The process is based on pulsed electrical currents delivered to a product placed between a set of electrodes; the distance between electrodes is termed as the treatment gap of the PEF chamber. The applied high voltage results in an electric field that causes microbial inactivation. The electric field may be applied in the form of exponentially decaying, square wave, bipolar, or oscillatory pulses and at ambient, sub-ambient, or slightly above-ambient temperature. After the treatment, the food is packaged aseptically and stored under refrigeration. applied to a food product held between two electrodes inside a chamber, usually at room temperature. Food is capable of transferring electricity because of the presence of several ions, giving the product in question a certain degree of electrical conductivity. So, when an electrical field is applied, electrical current flows into the liquid food and is transferred to each point in the liquid because of the charged molecules present.

Several nonthermal processing technologies were proposed on the basis of the same basic principle of keeping food below temperatures normally used in thermal processing. This would retain the nutritional quality of food including vitamins, minerals, and essential flavors while consuming less energy than thermal processing. High hydrostatic pressure, oscillating magnetic fields, intense light pulses, irradiation, the use of chemicals and biochemicals, high intensity pulse electric fields, and the hurdle concept were all recognized as emerging nonthermal technologies in recent years.

As a result of this permanent membrane damage, microorganisms are inactivated. Some applications of PEF technology are in biotechnology and genetic engineering for electroporation in cell hybridization.

Several nonthermal processing technologies were proposed on the basis of the same basic principle of keeping food below temperatures normally used in thermal processing. This would retain the nutritional quality of food including vitamins, minerals, and essential flavors while consuming less energy than thermal processing. High hydrostatic pressure, oscillating magnetic fields, intense light pulses, irradiation, the use of chemicals and biochemicals, high intensity pulse electric fields, and the hurdle concept were all recognized as emerging nonthermal technologies in recent years.

The basis for this prediction is because of PEF ability to inactivate microorganisms in the food, reduce enzymatic activity, and extend shelf-life with negligible changes in the quality of the final product as compared to the original one. According to the intensity of the field strength, electroporation can be either reversible (cell membrane discharge) or irreversible (cell membrane breakdown or lysis), but this effect can be controlled depending on the application.

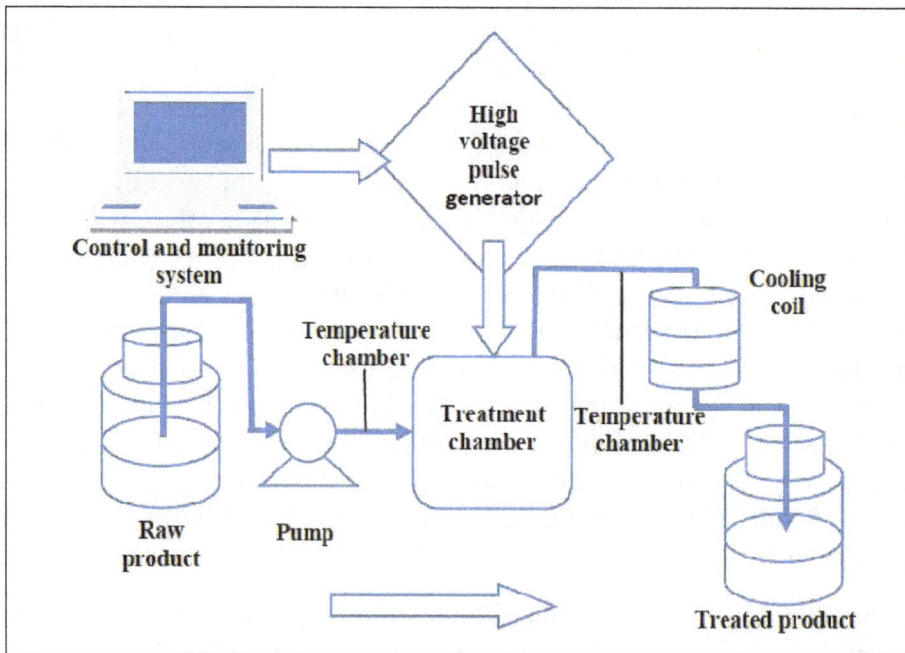

Flow chart of a PEF food processing system with basic component.

PEF technology is based on a pulsing power delivered to the product placed between a set of electrodes confining the treatment gap of the PEF chamber. The equipment consists of a high voltage pulse generator and a treatment chamber with a suitable fluid handling system and necessary monitoring and controlling devices. Food product is placed in the treatment chamber, either in a static or continuous design, where two electrodes are connected together with a nonconductive material to avoid electrical flow from one to the other. Generated high voltage electrical pulses are applied to the electrodes, which then conduct the high intensity electrical pulse to the product placed between the two electrodes. The food product experiences a force per unit charge, the so-called electric field, which is responsible for the irreversible cell membrane breakdown in microorganisms.

This leads to dielectric breakdown of the microbial cell membranes and to interaction with the charged molecules of food. Hence, PEF technology has been suggested for the pasteurization of foods such as juices, milk, yogurt, soups, and liquid eggs.

System Components

A pulsed Electric Field processing system consists of a high-voltage power source, an energy storage capacitor bank, a charging current limiting resistor, a switch to discharge energy from the capacitor across the food and a treatment chamber. An oscilloscope is used to observe the pulse waveform. The power source, a high voltage DC generator, converts voltage from an utility line (110 V) into high voltage AC, then rectifies to a high voltage DC. Energy from the power source is stored in the capacitor and is discharged through the treatment chamber to generate an electric field in the food material. The maximum voltage across the capacitor is equal to the voltage across the generator. The bank of capacitors is charged by a direct current power source obtained from amplified and rectified regular alternative current main source. An electrical switch is used to discharge energy (instantaneously in millionth of a second) stored in the capacitor storage bank across the food held in the treatment chamber. Apart from those major components, some adjunct parts are also necessary. In case of continuous systems, a pump is used to convey the food through the treatment chamber. A chamber cooling system may be used to diminish the ohmic heating effect and control food temperature during treatment. High-voltage and high-current probes are used to measure the voltage and current delivered to the chamber. Figure shows a basic PEF treatment unit.

A PEF system for food processing in general consists of three basic components: a high voltage pulse generator, a treatment chamber and a control system for monitoring the process parameters.

Many successful steps have been taken in the design of system components and inactivation mechanism for different species. Methods applied to thermal processing technologies by plotting logs of the numbers of survivors against log or treatment time, or number of pulses, have been used to explain inactivation kinetics neglecting the deviations from linearity for these plots.

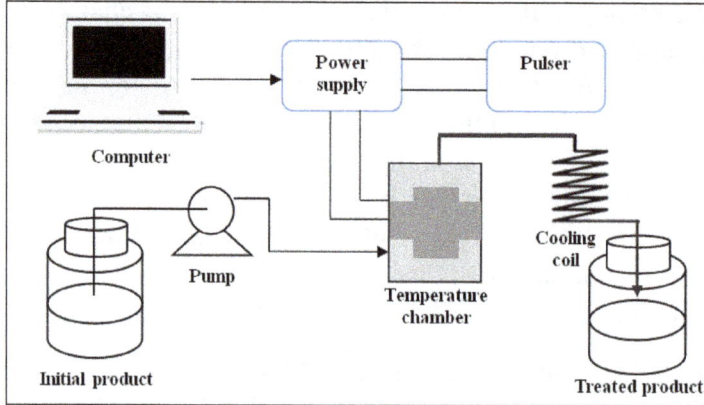

Schematic diagram of a pulsed electric fields operation.

The high intensity pulsed electric field processing system is a simple electrical system consisting of a high voltage source, capacitor bank, switch, and treatment chamber. Generation of pulsed electric fields requires a fast discharge of electrical energy within a short period of time. This is accomplished by the pulse-forming network (PFN), an electrical circuit consisting of one or more power supplies with the ability to charge voltages (up to 60 kV), switches (ignitron, thyratron, tetrode, spark gap, semiconductors), capacitors (0.1-10 µF), resistors (2Ω-1O MΩ), and treatment chambers.

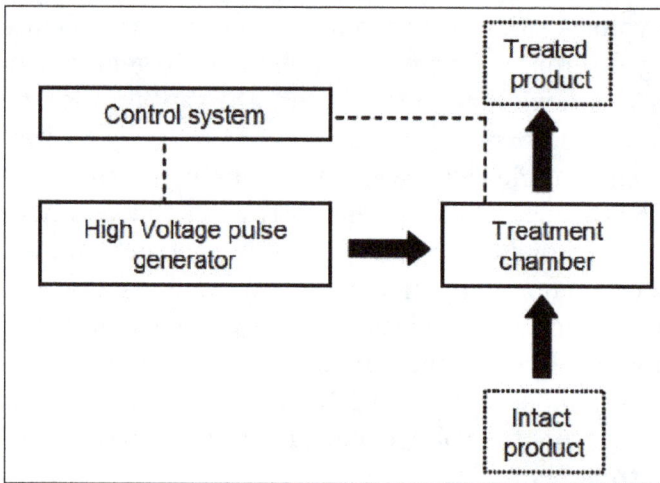

Scheme of a pulsed electric field system for food processing.

The PEF processing system is composed of a high voltage repetitive pulser, a treatment chamber(s), a cooling system(s), voltage and current measuring devices, a control unit, and a data acquisition system. A pulsed power supply is used to obtain high voltage from low utility level voltage, and the former is used to charge a capacitor bank and switch to discharge energy from the capacitor across the food in the treatment chamber. Treatment chambers are designed to hold the food during PEF processing and house the discharging electrodes. After processing the product is cooled, if necessary, packed aseptically, and then stored at refrigerated or ambient temperatures depending on the type of food.

Power Supply

High voltage pulses are supplied to the system via a high voltage pulse generator at required intensity, shape, and duration. The high voltage power supply for the system can either be an ordinary source of direct current (D) or a capacitor charging power supply with high frequency AC inputs that provide a command charge with higher repetitive rates than the DC power supply.

High voltage pulses are supplied to the PEF system via a high voltage generator at required electric field intensity, pulse waveform and pulse width. In general, the high voltage power supply is used to charge the capacitor bank and store the energy to the capacitor bank. Liquid food may be processed in a static treatment chamber or in a continuous treatment chamber through a pump. For preliminary laboratory-scale studies, the static treatment chamber is used, but a continuous treatment chamber is desirable for the pilot plant or industrial-scale operations. In order to avoid undesirable thermal effects, cold water of the cooling system is recirculated through the electrodes to dissipate the heat generated by the electric current passing through the food.

High-power Capacitors

The main components of high-power sources are storage capacitors and on- and off-switches. Because of their relatively high ohmic power consumption, inductors in comparison to capacitors play a minor role. The energy stored in capacitors is used to generate electric or magnetic fields. Electric fields are used to accelerate charged particles, leading to thermal, chemical, mechanical, electromagnetic wave, or breakdown effects. Electromagnetic fields transfer energy as electromagnetic waves, xray, microwaves, and laser beam generation are typical examples. Magnetic fields facilitate the generation of extremely high pressures ranging from 0.1 GPa to many GPa. These effects are applied to modify molecules to remodell, compress, weld, segment, fragment, or destroy materials; and to modify the surface of organic and inorganic parts and particles.

Switches

The discharging switch also plays a critical role in the efficiency of the PEF system. The type of switch used will determine how fast it can perform and how much current and voltage it can withstand. In increasing order of service life, suitable switches for PEF systems include: ignitrons, spark gaps, trigatrons, thyratrons, and semiconductors. Solid-state semiconductor switches are considered by the experts as the future of high power switching.

After the energy storage device, the switch is the most important element of a high-power pulse generator. High-power switching systems are the connecting elements between the storage device and the load. The rise time, shape, and amplitude of the generator output pulse depends strongly on the properties of the switches in the pulse forming elements. Generators with capacitive storage devices need closing switches, while generators with inductive storage devices require opening switches.

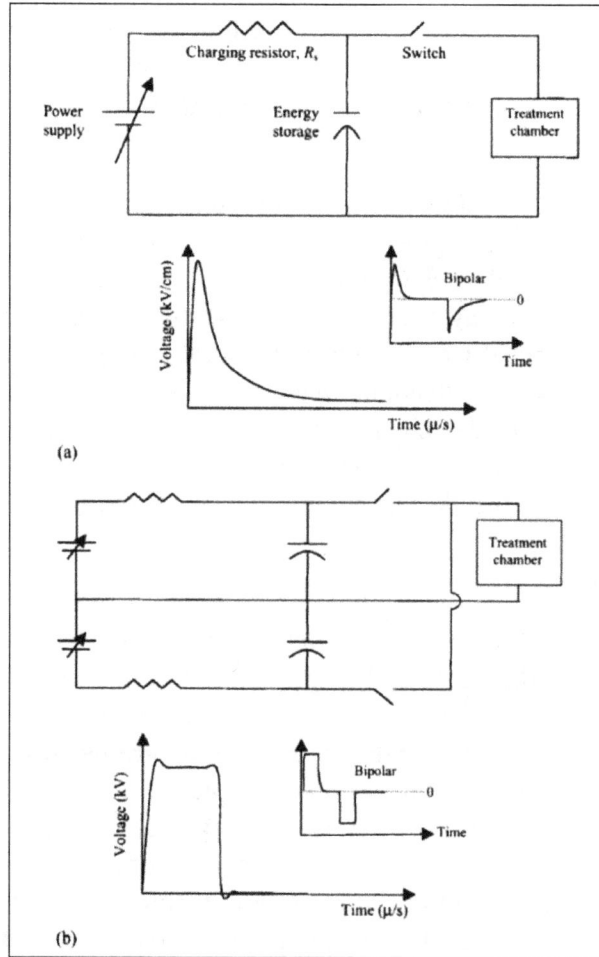

Commonly used pulse wave shapes and the generic electrical circuits: (a) Monopolar exponential decaying circuits and possible waveform; (b) Monopolar square circuit and possible waveform.

There are two main groups of switches currently available: ON switches and ON/OFF switches. ON switches provide full discharging of the capacitor but can only be turned off when discharging is completed. ON switches can handle high voltages with relatively lower cost compared to ON/OFF switches, however, the short life and low repetition rate are some disadvantages to be considered for selection. The Ignitron, Gas Spark Gap, Trigatron, and Thyratron are some of the examples from this group. ON/OFF type switches have been developed in recent years that provide control over the pulse generation process with partial or complete discharge of the capacitors. Improvements on switches, mainly on semiconductor solid-states witches, have resulted in longer life spans and better performance. The gate turn off (GTO) thyristor, the insulated gate bipolar transistor (IGBT), and the symmetrical gate commutated thyristor (SGCT) are some examples from this group.

High Voltage Pulse Generator

The high voltage pulse generator provides electrical pulses of the desired voltage, shape and duration by using a more or less complex pulse forming network (PFN). More in detail, a PFN is an electrical circuit consisting of several components: one or more DC power supplies, a charging resistor, a capacitor bank formed by two or more units connected in parallel, one or more switches, and pulse-shaping inductors and resistors. The DC power supply charges the capacitors bank to the desired voltage. Using this device, the ac power from the utility line (50-60Hz) is converted in high voltage alternating current power and then rectified to high voltage DC power.

A low-energy PEF system, which consists of a high voltage pulse generator is used to treat the spoiled grape juice samples. The details are given by Ho and Mittal. The system consists of a 30 kV d.c. high-voltage pulse generator, a circular treatment chamber, and devices for pumping and recording. The 110V a.c. was raised in voltage through a high-voltage transformer, and then rectified. The d.c. high-voltage supply then charges up the 0.12 uF capacitor through a series of 6 MΩ resistors (the time constant =0.72 s). The pulse generator emits a train of 5V pulses, and the trigger circuit serves to convert that to 500V pulses using a silicon control rectifier (SCR). The generation of high voltage pulses relies on the discharge of the 0.12 uF capacitor through the thyratron. The batch unit can generate short duration pulses (2 ms width, 0.5Hz frequency) with a peak-to-peak electric field strength up to 100 kV/cm. The uniqueness of this pulser is that the pulses of low energy (<25 J/pulse) and of instant charge reversal shape are generated.

Treatment Chamber

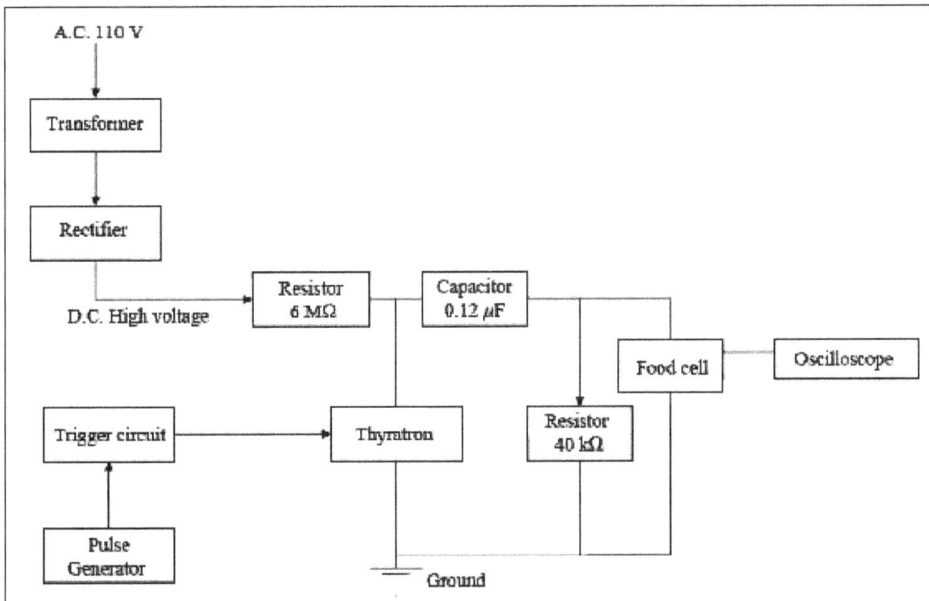

Generalized scheme of pulsed electric field equipment.

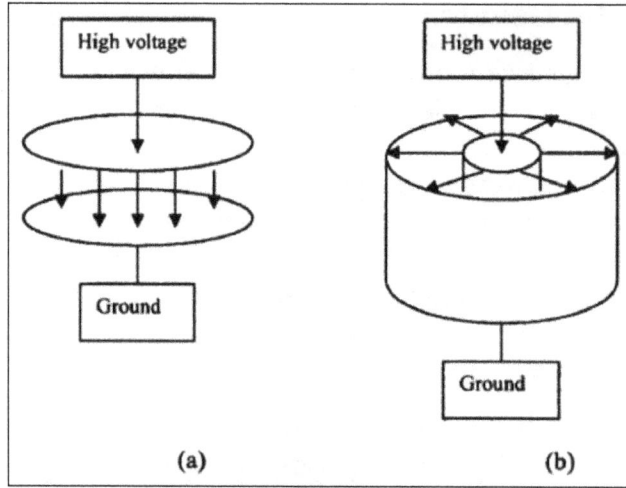

Common electrode configurations in pulsed electric
field treatment chambers: (a) parallel-plate, (b) coaxial.

One of the most important and complicated components in the processing system is the treatment chamber. The basic idea of the treatment chamber is to keep the treated product inside during pulsing, although the uniformity of the process is highly dependent on the characteristic design of the treatment chamber. When the strength of applied electric fields exceeds the electric field strength of the food product treated in the chamber, breakdown of food occurs as a spark. Treatment chambers are mainly grouped together to operate in either a batch or continuous manner; batch systems are generally found in early designs for handling of static volumes of solid or semi-solid foods. Several treatment chambers have been designed. they can be categorized within two types: parallel plate and coaxial. Parallel plate chambers have been typically used in batch modes while coaxial designs have been used in continuous modes where the medium is pumped through at a known flow rate and pulses are applied at a known pulse frequency. Coaxial chambers used in continuous operation have been found to result in higher inactivation rates compared to batch systems sins there is a more uniform diatribution of the electric field in continuously flowing media presents different chamber designs.

PEF investigators studying inactivation and preservation effects have been highly inventive in treatment chamber design. Several different designs have been developed through the years for this key component, wherein high voltage delivered by the power supply is applied to the product located between a pair of electrodes. The basic idea of the treatment chamber is to keep the treated product inside during pulsing, although the uniformity of the process is highly dependent on the characteristic design of the treatment chamber. When the strength of applied electric fields exceeds the electric field strength of the food product treated in the chamber, break down of food occurs as a spark. Known as the dielectric breakdown of food, this is one of the most important concepts to be considered in treatment chamber design. Dielectric breakdown of the food is generally characterized as causing damage on the electrode surfaces in the form of pits, a result of arching and increased pressure, leading to treatment chamber explosions and evolution of gas bubbles. Intrinsic electrical resistance, electric field homogeneity, and reduction and generation of enhanced field areas are some other important design criteria for a successful design in terms of energy consumption and low product heating.

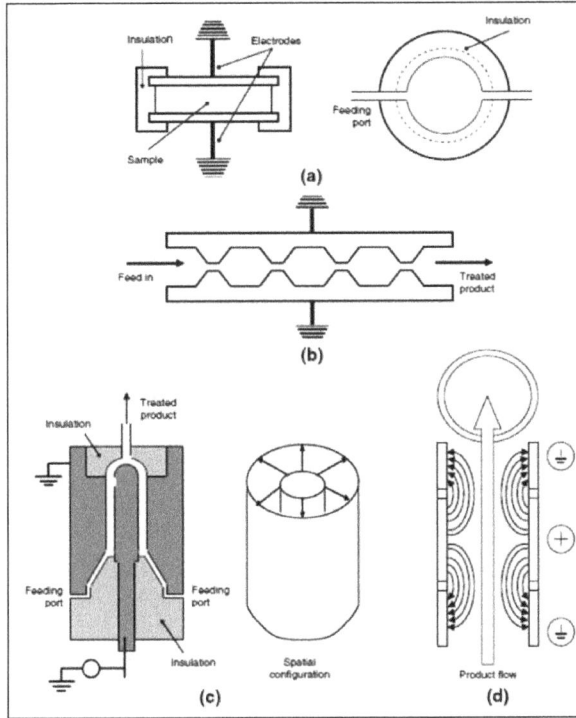

Schematic diagram of a pulsed electric fields operation design of treatment chambers for pulsed electric fields equipment: a) static chamber, b) side view of a basic continuous design, c) coaxial chamber, and d) co-linear chamber.

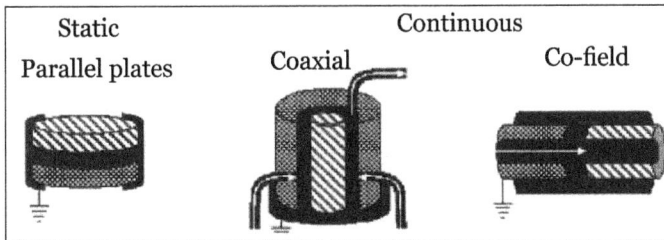

Schematic configurations of the three most used PEF treatment chambers.

Dunn and Pearlman designed a chamber consisting of two parallel plate electrodes and a dielectric space insulator. The electrodes are separated from the food by ion conductive membranes made of sulfonated polystyrene and acrylic acid copolymers, but fluorinated hydrocarbon polymers with pendant groups would also be suitable. An electrolyte is used to facilitate electrical conduction between electrodes and ion permeable membranes. Suitable electrolyte solutions include sodium carbonate, sodium hydroxide, potassium carbonate, and potassium hydroxide. These are circulated continuously to remove the products of electrolysis and replaced in the event of excess concentration or depletion of ionic components.

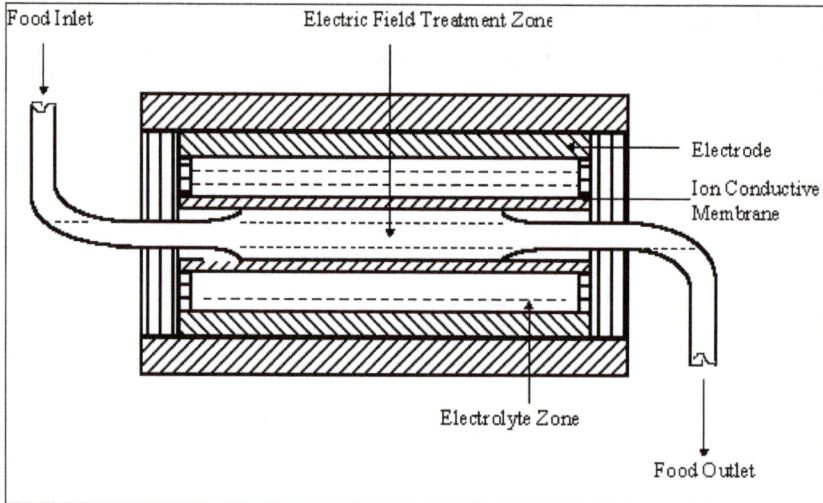

A continuous chamber with ion-conductive membranes separating the electrodes and food.

From the electrical point of view, the PEF treatment chamber represents the electrical load consisting of two or more electrodes filled with the liquid substance to be treated. The chamber has to be constructed in such a way that the electrical field acting on the liquid is more or less homogeneous across the entire active region. Planar electrode configurations consist of two parallel electrodes fixed by insulators. The insulators and the electrodes form a channel for the streaming liquid. Coaxial electrode configurations consist of two coaxial electrodes. The liquid streams between these electrodes that are fixed by insulators. Axial electrode configurations consist of several electrode rings on alternating potentials separated by insulating rings. Electrode materials also play an essential role. If monopolar voltage waveforms are applied, electrode corrosion can become critical and the substance to be treated can be contaminated. In commercially available electroporation devices with small probes, aluminum, stainless steel, carbon, gold-plated electrodes, and even silver electrodes are used.

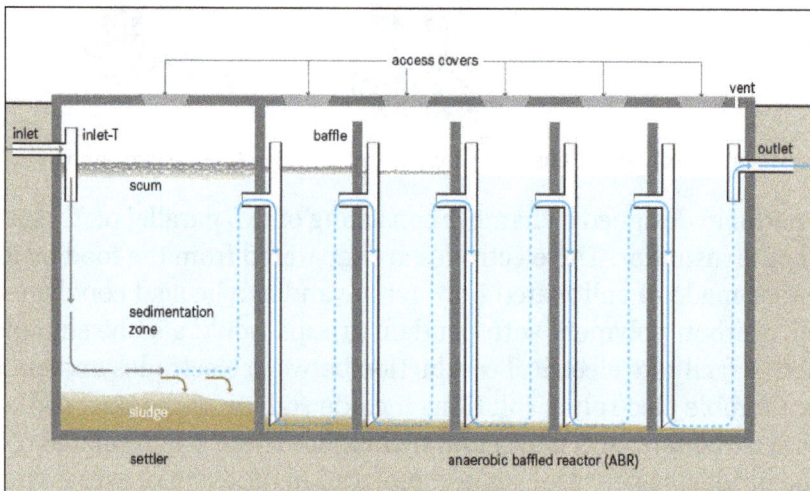

Anaerobic baffles chamber.

The static parallel plate electrode chamber was modified by adding baffled flow channels inside to make it operate as a continuous chamber. Two stainless-steel disk-shaped electrodes separated by a polysulfone spacer form the chamber. The designed operating conditions are: chamber volume, 20 or 8 ml; electrode gap, 0.95 or 0.51 cm; and food flow rate, 1200 or 6 ml/min.

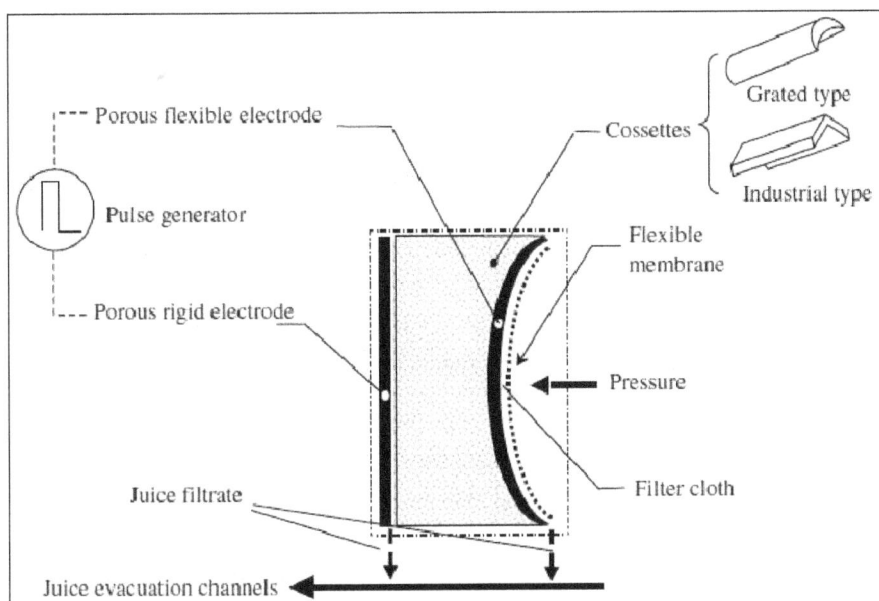

Schematic of a one chamber configuration of plate and frame filter press.

In most runs, one or two chambers were manually filled with grated cossettes. In a few runs, up to six chambers were used at the same time. Each chamber consists of a plate covered by a filter cloth and a flexible electrode (metallic grid) on one side and a rigid electrode on the other. The pressure (compressed air) is applied to the membrane of the plate, which in turn exerts and distributes the pressure over the cossettes placed between the plate and the rigid electrode. Juice is drained through channels leading to the outlet, where juice accumulation is monitored by a weighing balance connected to a data acquisition system.

Applications of Pulsed Electric Fields Technology

Application of pulsed electric fields technology has been successfully demonstrated for the pasteurization of foods such as juices, milk, yogurt, soups, and liquid eggs. Application of PEF processing is restricted to food products with no air bubbles and with low electrical conductivity. The maximum particle size in the liquid must be smaller than the gap of the treatment region in the chamber in order to ensure proper treatment. PEF is a continuous processing method, which is not suitable for solid food products that are not pump able. PEF is also applied to enhance extraction of sugars and other cellular content from plant cells, such as sugar beets. PEF also found application in reducing the solid volume (sludge) of wastewater.

PEF processing has been successful in a variety of fruit juices with low viscosity and electrical conductivity such as orange, apple, and cranberry juice. Recent studies reported more than a 3-10 g reduction in orange juice and apple juice.

Additionally, the color change in fruit juices (subject to prolonged storage) was reportedly less in juices treated by PEF, as in a recent study of PEF-treated orange juice stored at 4 °C for 112 days; there was less browning than thermally pasteurized juice, which was attributed to conversion of ascorbic acid to furfural.

Considering the effectiveness of PEF treatment on liquid products, such as milk, fruit juices, liquid egg, and any other pumpable food products, extensive research has been done to implement the process at an industrial level. Flavor freshness, economic feasibility, improvements in functional and textural attributes and extended shelf life are some of the main points of interest besides achievement of microbiological safety of food products. Among all liquid products, PEF technology has been most widely applied to apple juice, orange juice, milk, liquid egg, and brine solutions.

Each of the nonthermal technologies has specific applications in terms of the types of foods that can be processed. Among these, pulsed electric fields (PEF) is one of the most promising nonthermal processing methods for inactivation of microorganisms, with the potential of being an alternative for pasteurization of liquid foods. Comparable to pasteurization, yet without the thermal component, PEF has the potential to pasteurize several foods via exposure to high voltage short pulses maintained at temperatures below 30-40 °C. The basic definition of PEF technology relies on the use of high intensity pulsed electric fields (lo-80 kV/cm) for cell membrane disruption where induced electric fields perforate microbial membranes by electroporation, a biotechnology process used to promote bacterial DNA interchange. Induction of membrane potentials exceeding a threshold value often result in cell damage and death.

PEF technology has recently been used in alternative applications including drying enhancement, enzyme activity modification, preservation of solid and semisolid food products, and waste water treatment, besides pretreatment applications for improvement of metabolite extraction. The ability of PEF to increase permeabilization means it can be successfully used to enhance mass and heat transfer to assist drying of plant tissues.

Considering the effectiveness of PEF treatment on liquid products, such as milk, fruit juices, liquid egg, and any other pump able food products, extensive research has been done to implement the process at an industrial level. Flavor freshness, economic feasibility, improvements in functional and textural attributes and extended shelf life are some of the main points of interest besides achievement of microbiological safety of food products.

Among all liquid products, PEF technology has been most widely applied to apple juice, orange juice, milk, liquid egg, and brine solutions.

Application of PEF is especially promising for the citrus industry, which is concerned with the spoilage microorganisms and resultant production of off-flavor compounds such as lactic acid bacteria.

Jemai and Vorobiev stated the enhancing effect of a PEF treatment on the diffusion co-efficients of soluble substances in apple slices. The results available in literature clearly indicate that PEF can be successfully applied to disintegrate biological tissue and to improve the release of intracellular compounds, though an industrial application has not been achieved up to now. At Berlin University of Technology a system with a peak voltage of 20 kV, an average power of 7 kW and a production capacity of 2 ton/h has been developed for the treatment of fruit mashes. The power supply and the treatment chamber are shown in figure. It is noteworthy that avoiding an enzymatic maceration the pectin fractions will remain in a native, highly esterified structure. This provides a potential to extract high quality pectin from the pomace after juice winning and therefore a step toward a more economic and sustainable processing.

PEF treatments are applied in the form of short pulses to avoid excessive heating or undesirable electrolytic reactions. In general, a continuous PEF treatment system is composed of treatment chambers, a pulse generator, a fluid-handling system, and monitoring systems.

An OSU-4D bench-scale continuous unit manufactured in Ohio State University (US) was used to treat the food sample. Six co-field chambers with a diameter of 2. 3×10^{-3} m and a gap distance of 2.93×10^{-3} m between electrodes were connected in series. Two cooling coils were connected before and after each pair of chambers and submerged in a circulating bath to maintain the selected temperature at 35 or 55 °C. The temperature was recorded by thermocouples (T type, ±0.10) at the entrance of the first treatment chamber (initial temperature) and at the edcxit of the last treatment chamber (final temperature).

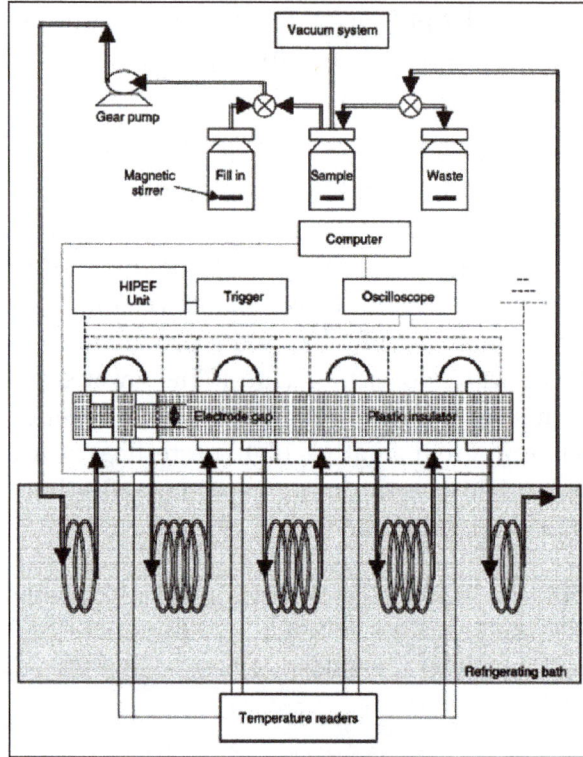

Diagram of the high-intensity pulsed electric field bench-scale processing unit.

Vorobiev, developed a laboratory device in the University of Technology of Compiegne (UT) permits both pretreatment and intermediate treatment by PEF. The treatment cell has a polypropylene frame with a cylindrical cavity compartment (20 mm thick, 56 mm in diameter), which should be initially filled with gratings and then closed from both sides by steel covers. A mobile electrode is attached to the elastic rubber diaphragm. A stationary wire gauze electrode is installed between the filter cloth and the layer of gratings. Both electrodes are connected to the PEF generator, which can provide the monopolar or bipolar pulses of near-rectangular shape.

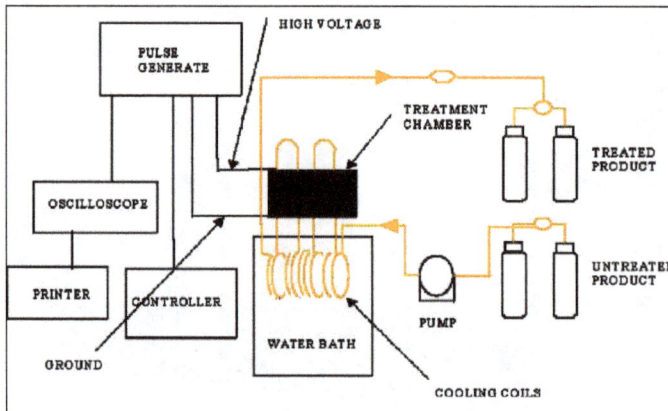

Scheme of the pulsed electric field equipment.

Studies conducted on the effects of PEF on dairy products such as skim milk, whole milk, and yogurt compromise a major section of PEF applications. Milk is very susceptible to both spoilage and pathogenic microorganisms requiring the application of thermal pasteurization under current regulations, which ensures safety but generally results in a cooked flavour.

Factors Affecting the Outcome of Pulsed Electric Fields Treatments

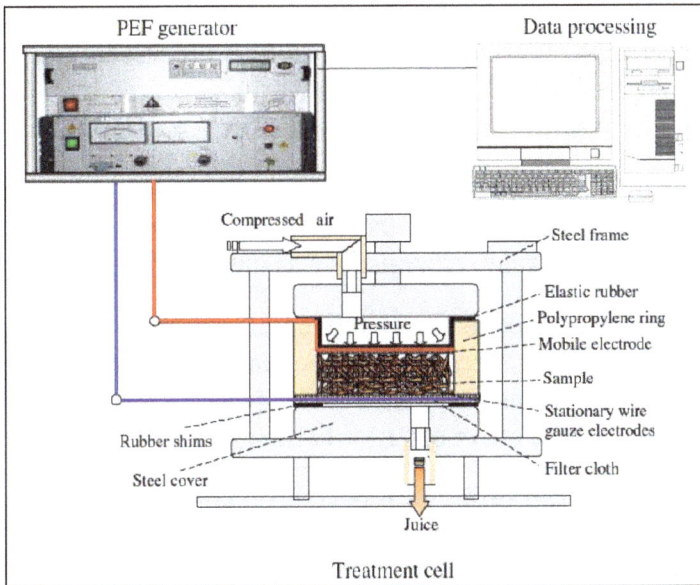

Experimental setup for permits both pretreatment and intermediate treatment by PEF.

In order to use PEF technology as a pasteurization process it is necessary to estimate its efficacy against pathogenic and spoilage food-borne microorganisms. To obtain this objective there is a need to accumulate knowledge on the critical factors affecting microbial inactivation, to describe the PEF inactivation kinetics and to understand the mechanisms involved in microbial PEF inactivation. The lethality factors contributing to the effectiveness of pulsed electric field technology can be grouped as technological, biological, and media factors. Each group of determinant factors is related to type of equipment, processing parameters, target microorganism, and type and condition of media used.

Technological Factors

A number of other factors during PEF processing can affect specific microbial inactivation as well. Some of these critical factors include the field strength, treatment time, treatment temperature, pulse shape, type of microorganism, growth stage of microorganism, and characteristics of the treatment substrate.

Microbial inactivation increases with an increase in the electric field intensity, above a critical trans membrane potential.

It is important that the electric field intensity should be evenly distributed in the treatment chamber to achieve an efficient treatment. Electric field intensities of smaller than 4-8 kV cm^{-1} usually do not affect microbial inactivation.

In general, the electric field intensity required to inactivate microorganisms in foods in the range of 12-45 kV cm^{-1}. The fact that microbial inactivation increases with increases in the applied electric field intensity and can be attributed to the high energy supplied to the cell suspension in a liquid product.

An important aspect that differentiates between PEF processing and other microbial inactivation technologies is that the PEF treatment is delivered by pulsing. The pulses commonly used in PEF treatments are usually either exponential or square wave pulses.

There is some controversy with respect to the influence of the pulse width on the PEF microbial lethality. Some authors have indicated that after the same treatment time, inactivation tested in several microorganisms was independent of the pulse width.

Treatment time could be defined as the effective time during which range microorganisms are subjected to the field strength. It depends on the number of pulses and the width of the pulses applied. This parameter and the electric field strength are the main factors determining the lethal effect of PEF treatments.

Studies on microbial inactivation by PEF have been conducted at frequencies ranged from 1 to 500 Hz. If the same number of pulses is applied, microbial inactivation is generally independent of the number of pulses applied per second.

PEF treatment time is calculated by multiplying the pulse number by the pulse duration. An increase in any of these variables increases microbial inactivation.

A good understanding of the electrical principles behind PEF technology is essential for a comprehensive analysis of the PEF system. The electrical field concept, introduced by Faraday, explains the electrical field force acting between two charges. When unit positive charge q located at a certain point within the electric field is generated in the treatment gap (Er), it experiences force F identified by position vector r. The electrical field per unit charge is then defined as shown in equation.:

$$E_r = \frac{F_{qr}}{q}$$

The electrical potential difference (V) between voltage across two points, separated by a nonconductive material, results in generation of an electric field between these points, with an electrical intensity (E) directly proportional to the magnitude of potential difference (V) and inversely proportional to the distance (d) between points, as given in equation:

$$E = \frac{V}{d}$$

The type of electrical field waveform applied is one of the important descriptive characteristics of a pulsed electric field treatment system. The exponentially decaying or square waves are among the most common waveforms used. To generate an exponentially decaying voltage wave, a DC power supply charges the bank of capacitors that are connected in series with a charging resistor. When a trigger signal is applied, the charge stored in the capacitor flows through the food in the treatment chamber. Exponential waveforms are easier to generate from the generator point of view. Generation of square waveform generally requires a pulse forming network (PFN) consisting of an array of capacitors and inductors. It is more challenging to design a square waveform system compared to an exponential waveform system. However, square waveforms may be more lethal and energy efficient than exponentially decaying pulses since square pulses have longer peak voltage duration compared to exponential pulses.

The electric field should be evenly distributed in the treatment chamber in order to achieve an efficient treatment. An electric field intensity of 16 kV/cm or greater is usually sufficient to reduce the viability of Gram negative bacteria by 4 to 5 log cycles and Gram positive bacteria by 3 to 4 log cycles.

Biological Factors

Biological factors that include the individual characteristics of target microorganisms and their physiological and growth states are determinant factors affecting PEF application. The susceptibility of a microorganism to PEF inactivation is highly related to the intrinsic parameters of the microorganism such as size, shape, species or growth state. Generally, Gram-positive vegetative cells are more resistant to PEF than Gram-negative bacteria, while yeasts show a higher sensitivity than bacteria. Induction of electric fields into cell membranes is greater when larger cells are exposed to PEF treatment. Most of the research focuses on the inactivation of vegetative cells of bacteria, while only a few reports are available on the inactivation of spores describing a limited effect of PEF. Bacillus cereus spores were mostly resistant (approximately 1 log reduction) to a mild PEF treatment at electric field strength of 20 kV/cm and 10.4 pulses in a study conducted on apple juice.

Another study conducted by Pagan et al. found that Bacillus cereus spores were not affected with PEF treatment of 60 kV/cm for 75 pulses at room temperature. On the other hand, Marquez et al. reported 3.42-log and 5-log reductions of Bacillus subtilis and Bacillus cereus spores, respectively, with PEF treatment of 50 kV/cm for 50 pulses at 25 °C in salt solution. Additionally, mold Condispores were reported to be sensitive to PEF in fruit juices whereas Neosartorya fischeriasco spores were resistant to PEF treatments.

Compared to the number of studies reported for enzyme inactivation by PEF, little information is available on the mechanism of inactivation, which may be due to the lack of analysis of enzyme structural data.

Media Factors

The effects of PEF on the food system are related to the PEF system and the properties of the liquid food. The most important factors in the PEF system are the electric field intensity, number of pulses, pulse waveform, pulse width, treatment time and treatment temperature. But enzymes and proteins are generally more resistant to electric field intensity and pulses than microorganisms. This requires further investigation, especially on the effects of pH, temperature, resistivity and composition of the enzyme or protein-containing medium or food system.

The physical and chemical characteristics of food products are known to strongly influence the effectiveness of microbial inactivation during PEF application, thus the challenge experienced using real food systems was due to the important role of the media's chemical and physical characteristics. These factors most likely influence the recovery of injured microbial cells and their subsequent growth following PEF exposure, since the presence of food components, such as fats and proteins, has reportedly had a preventive effect on microorganisms against PEF treatment.

Similar to the intrinsic parameters of microorganisms, treated media has its own intrinsic factor s such as conductivity, resistivity, dielectric properties, ionic strength, pH, and composition. Each of these parameter sinfluences the PEF treatment either alone or in combination. PEF technology has recently been used in alternative applications including drying enhancement, enzyme activity modification, preservation of solid and semisolid food products, and waste water treatment, besides pretreatment applications for improvement of metabolite extraction. The ability of PEF to increase permeabilization means it can be successfully used to enhance mass and heat transfer to assist drying of plant tissues. reported increased yield of water removal by 20-30% when exposed to low intensity electric fields.

Temperature is one factor proposed that has been correlated with microbial inactivation, and although PEF application is strictly a nonthermal processing technology, the synergistic effect of temperature on foods (due to changes in the properties of cell membranes) becomes greater when foods are subjected to high intensity pulse electric fields. In general, the lethality of PEF treatments increases with an increase in processing temperature; therefore, a proper cooling device is necessary to maintain temperatures below levels. that affect nutritional, sensory or functional properties of food products.

Sepulveda proposed that a PEF treatment time between 0.1 to 0.5 ms produced the best results for microbial inactivation. The pulse width is defined as the time where the peak field is maintained for square wave pulses or the time until decay to 37% for exponential decay pulses. Typically, increasing the number of pulses causes an increase in treatment time, as the pulse width is fixed by the impulse generation setup.

Dunn and Pearlman found that a combination of PEF and heat was more efficient than conventional heat treatment alone. A higher level of inactivation was obtained using a combination of 55 °C temperature and PEF to treat milk.

Zhang et al. reported that increasing treatment temperature from 7 to 20 °C significantly increased PEF inactivation of E. coli in simulated milk ultra-filtrate (SMUF). However, additional increase in temperature from 20 to 33 °C did not result in any further increase in PEF inactivation.

References

- What-is-food-engineering: whatisengineering.com, Retrieved 31 March, 2019
- Food-Technology, Engineering-branches: myklassroom.com, Retrieved 14 July, 2019
- Food-processing: newworldencyclopedia.org, Retrieved 17 May, 2019
- Pulsed-electric-fields-for-food-processing-technology, structure-and-function-of-food-engineering: intechopen.com, Retrieved 19 April, 2019

Food Preservation

Food preservation is the process of treating and handling food in such a way that it minimizes the possibility of foodborne illnesses. Some of the diverse methods of food preservation are food drying, pickling, fermentation, food irradiation, brining and smoking. This chapter discusses in detail these methods of food preservation.

Food preservation is a number of methods by which food is kept from spoilage after harvest or slaughter. Such practices date to prehistoric times. Among the oldest methods of preservation are drying, refrigeration, and fermentation. Modern methods include canning, pasteurization, freezing, irradiation, and the addition of chemicals. Advances in packaging materials have played an important role in modern food preservation.

Spoilage Mechanisms

Food spoilage may be defined as any change that renders food unfit for human consumption. These changes may be caused by various factors, including contamination by microorganisms, infestation by insects, or degradation by endogenous enzymes (those present naturally in the food). In addition, physical and chemical changes, such as the tearing of plant or animal tissues or the oxidation of certain constituents of food, may promote food spoilage. Foods obtained from plant or animal sources begin to spoil soon after harvest or slaughter. The enzymes contained in the cells of plant and animal tissues may be released as a result of any mechanical damage inflicted during postharvest handling. These enzymes begin to break down the cellular material. The chemical reactions catalyzed by the enzymes result in the degradation of food quality, such as the development of off-flavours, the deterioration of texture, and the loss of nutrients. The typical microorganisms that cause food spoilage are bacteria (e.g., Lactobacillus), yeasts (e.g., Saccharomyces), and molds (e.g., Rhizopus).

Microbial Contamination

Bacteria and fungi (yeasts and molds) are the principal types of microorganisms that cause food spoilage and food-borne illnesses. Foods may be contaminated by microorganisms at any time during harvest, storage, processing, distribution, handling, or preparation. The primary sources of microbial contamination are soil, air, animal feed, animal hides and intestines, plant surfaces, sewage, and food processing machinery or utensils.

Bacteria

Bacteria are unicellular organisms that have a simple internal structure compared with the cells of other organisms. The increase in the number of bacteria in a population is commonly referred to as bacterial growth by microbiologists. This growth is the result of the division of one bacterial cell into two identical bacterial cells, a process called binary fission. Under optimal growth conditions, a bacterial cell may divide approximately every 20 minutes. Thus, a single cell can produce almost 70 billion cells in 12 hours. The factors that influence the growth of bacteria include nutrient availability, moisture, pH, oxygen levels, and the presence or absence of inhibiting substances (e.g., antibiotics).

The nutritional requirements of most bacteria are chemical elements such as carbon, hydrogen, oxygen, nitrogen, phosphorus, sulfur, magnesium, potassium, sodium, calcium, and iron. The bacteria obtain these elements by utilizing gases in the atmosphere and by metabolizing certain food constituents such as carbohydrates and proteins.

Temperature and pH play a significant role in controlling the growth rates of bacteria. Bacteria may be classified into three groups based on their temperature requirement for optimal growth: thermophiles (55–75 °C, or 130–170 °F), mesophiles (20–45 °C, or 70–115 °F), or psychrotrophs (10–20 °C, or 50–70 °F). In addition, most bacteria grow best in a neutral environment (pH equal to 7).

Bacteria also require a certain amount of available water for their growth. The availability of water is expressed as water activity and is defined by the ratio of the vapour pressure of water in the food to the vapour pressure of pure water at a specific temperature. Therefore, the water activity of any food product is always a value between 0 and 1, with 0 representing an absence of water and 1 representing pure water. Most bacteria do not grow in foods with a water activity below 0.91, although some halophilic bacteria (those able to tolerate high salt concentrations) can grow in foods with a water activity lower than 0.75. Growth may be controlled by lowering the water activity—either by adding solutes such as sugar, glycerol, and salt or by removing water through dehydration.

The oxygen requirements for optimal growth vary considerably for different bacteria. Some bacteria require the presence of free oxygen for growth and are called obligate aerobes, whereas other bacteria are poisoned by the presence of oxygen and are called obligate anaerobes. Facultative anaerobes are bacteria that can grow in both the presence or absence of oxygen. In addition to oxygen concentration, the oxygen reduction potential of the growth medium influences bacterial growth. The oxygen reduction potential is a relative measure of the oxidizing or reducing capacity of the growth medium.

When bacteria contaminate a food substrate, it takes some time before they start growing. This lag phase is the period when the bacteria are adjusting to the environment. Following the lag phase is the log phase, in which population grows in a logarithmic fashion. As the population grows, the bacteria consume available nutrients and produce waste products. When the nutrient supply is depleted, the growth rate enters a stationary phase in which the number of viable bacteria cells remains the same. During the stationary phase, the rate of bacterial cell growth is equal to the rate of bacterial cell death. When the rate of cell death becomes greater than the rate of cell growth, the population enters the decline phase.

A bacterial population is expressed either per gram or per square centimetre of surface area. Rarely does the total bacterial population exceed 1010 cells per gram. A population of less than 106 cells per gram does not cause any noticeable spoilage except in raw milk. Populations of between 106 and 107 cells per gram cause spoilage in some foods; for example, they can generate off-odours in vacuum-packaged meats. Populations of between 107 and 108 cells per gram produce off-odours in meats and some vegetables. At levels above 5×107 cells per gram, most foods exhibit some form of spoilage.

When the conditions for bacterial cell growth are unfavourable (e.g., low or high temperatures or low moisture content), several species of bacteria can produce resistant cells called endospores. Endospores are highly resistant to heat, chemicals, desiccation (drying out), and ultraviolet light. The endospores may remain dormant for long periods of time. When conditions become favourable for growth (e.g., thawing of meats), the endospores germinate and produce viable cells that can begin exponential growth.

Fungi

The two types of fungi that are important in food spoilage are yeasts and molds. Molds are multicellular fungi that reproduce by the formation of spores (single cells that can grow into a mature fungus). Spores are formed in large numbers and are easily dispersed through the air. Once these spores land on a food substrate, they can grow and reproduce if conditions are favourable. Yeasts are unicellular fungi that are much larger than bacterial cells. They reproduce by cell division (binary fission) or budding.

The conditions affecting the growth of fungi are similar to those affecting bacteria. Both yeasts and molds are able to grow in an acidic environment (pH less than 7). The pH range for yeast growth is 3.5 to 4.5 and for molds is 3.5 to 8.0. The low pH of fruits is generally unfavourable for the growth of bacteria, but yeasts and molds can grow and cause spoilage in fruits. For example, species of the fungal genus Colletotrichum cause crown rot in bananas. Yeasts promote fermentation in fruits by breaking down sugars into alcohol and carbon dioxide. The amount of available water in a food product is also critical for the growth of fungi. Yeasts are unable to grow at a water activity of less than 0.9, and molds are unable to grow at a water activity below 0.8.

Control of Microbial Contamination

The most common methods used either to kill or to reduce the growth of microorganisms are the application of heat, the removal of water, the lowering of temperature during storage, the reduction of pH, the control of oxygen and carbon dioxide concentrations, and the removal of the nutrients needed for growth. The use of chemicals as preservatives is strictly regulated by governmental agencies such as the Food and Drug Administration (FDA) in the United States. Although a chemical may have preservative functions, its safety must be proved before it may be used in food products. To suppress yeast and mold growth in foods, a number of chemical preservatives are permitted. In the United States, the list of such chemicals, known as GRAS (Generally Recognized as Safe), includes compounds such as benzoic acid, sodium benzoate, propionic acid, sorbic acid, and sodium diacetate.

Chemical Deterioration

Enzymatic Reactions

Enzymes are large protein molecules that act as biological catalysts, accelerating chemical reactions without being consumed to any appreciable extent themselves. The activity of enzymes is specific for a certain set of chemical substrates, and it is dependent on both pH and temperature.

The living tissues of plants and animals maintain a balance of enzymatic activity. This balance is disrupted upon harvest or slaughter. In some cases, enzymes that play a useful role in living tissues may catalyze spoilage reactions following harvest or slaughter. For example, the enzyme pepsin is found in the stomach of all animals and is involved in the breakdown of proteins during the normal digestion process. However, soon after the slaughter of an animal, pepsin begins to break down the proteins of the organs, weakening the tissues and making them more susceptible to microbial contamination. After the harvesting of fruits, certain enzymes remain active within the cells of the plant tissues. These enzymes continue to catalyze the biochemical processes of ripening and may eventually lead to rotting, as can be observed in bananas. In addition, oxidative enzymes in fruits continue to carry out cellular respiration (the process of using oxygen to metabolize glucose for energy). This continued respiration decreases the shelf life of fresh fruits and may lead to spoilage. Respiration may be controlled by refrigerated storage or modified-atmosphere packaging. Table lists a number of enzymes involved in the degradation of food quality.

Enzymes that Cause Food Spoilage		
Enzyme	Food	Spoilage action
Ascorbic acid oxidase	Vegetables	Destruction of vitamin C
Lipase	Cereals	Discoloration
	Milk	Hydrolytic rancidity
	Oils	Hydrolytic rancidity

Lipoxygenase	Vegetables	Destruction of vitamin a, off-flavour
Pectic enzyme	Citrus juices	Destruction of pectic substances
	Fruits	Excessive softening
Peroxidase	Fruits	Browning
Polyphenoloxidase	Fruits, vegetables	Browning, off-flavour, vitamin loss
Protease	Eggs	Reduction of shelf life of fresh and dried whole eggs
	Crab, lobster	Overtenderization
	Flour	Reduction of gluten formation
Thiaminase	Meats, fish	Destruction of thiamine

Autoxidation

The unsaturated fatty acids present in the lipids of many foods are susceptible to chemical breakdown when exposed to oxygen. The oxidation of unsaturated fatty acids is autocatalytic; that is, it proceeds by a free-radical chain reaction. Free radicals contain an unpaired electron (represented by a dot in the molecular formula) and therefore, are highly reactive chemical molecules. The basic mechanisms in a free-radical chain reaction involve initiation, propagation, and termination steps. Under certain conditions, in initiation a free-radical molecule ($X \cdot$) present in the food removes a hydrogen (H) atom from a lipid molecule, producing a lipid radical ($L \cdot$). This lipid radical reacts with molecular oxygen (O_2) to form a peroxy radical ($LOO \cdot$). The peroxy radical removes a hydrogen atom from another lipid molecule and the reaction starts over again (propagation). During the propagation steps, hydroperoxide molecules (LOOH) are formed that may break down into alkoxy ($LO \cdot$) and peroxy radicals plus water (H_2O). The lipid, alkoxy, and peroxy radicals may combine with one another (or other radicals) to form stable, nonpropagating products (termination). These products result in the development of rancid off-flavours. In addition to promoting rancidity, the free radicals and peroxides produced in these reactions may have other negative effects, such as the bleaching of food colour and the destruction of vitamins A, C, and E. This type of deterioration is prevalent in fried snacks, nuts, cooking oils, and margarine.

The autoxidation of unsaturated fatty acids:

- Initiation: $X \cdot + LH \rightarrow L \cdot + XH$

  $L \cdot + O_2 \rightarrow LOO \cdot$
- Propagation: $LOO \cdot LH \rightarrow LOOH + L \cdot$

  $2LOOH \rightarrow LO \cdot + LOO \cdot + H_2O$
- Termination: $L \cdot, LO \cdot, LOO \cdot \rightarrow$ a number of stable nonpropagating species

Maillard Reaction

Another chemical reaction that causes major food spoilage is nonenzymatic browning, also known as the Maillard reaction. This reaction takes place between reducing sugars (simple monosaccharides capable of carrying out reduction reactions) and the amino group of proteins or amino acids present in foods. The products of the Maillard reaction lead to a darkening of colour, reduced solubility of proteins, development of bitter flavours, and reduced nutritional availability of certain amino acids such as lysine. The rate of this reaction is influenced by the water activity, temperature, and pH of the food product. Nonenzymatic browning causes spoilage during the storage of dry milk, dry whole eggs, and breakfast cereals.

Light-induced Reactions

Light influences a number of chemical reactions that lead to spoilage of foods. These light-induced reactions include the destruction of chlorophyll (the photosynthetic pigment that gives plants their green colour), resulting in the bleaching of certain vegetables; the discoloration of fresh meats; the destruction of riboflavin in milk; and the oxidation of vitamin C and carotenoid pigments (a process called photosensitized oxidation). The use of packaging material that prevents exposure to light is one of the most effective means of preventing light-induced chemical spoilage.

FOOD PRESERVATIVES

All food products except for the one growing in your kitchen garden has food preservatives in them. Every manufacturer adds food preservative to the food during processing. The purpose is generally to avoid spoilage during the transportation time.

Food is so important for the survival, so food preservation is one of the oldest technologies used by human beings to avoid its spoilage. Different ways and means have been found and improved for the purpose. Boiling, freezing & refrigeration, pasteurizing, dehydrating, pickling are the traditional few. Sugar, mineral salt and salt are also often used as preservatives food. Nuclear radiation is also being used now as food preservatives. Modified packaging techniques like vacuum packing and hypobaric packing also work as food preservatives.

Food preservation is basically done for three reasons:

- To preserve the natural characteristics of food.
- To preserve the appearance of food.
- To increase the shelf value of food for storage.

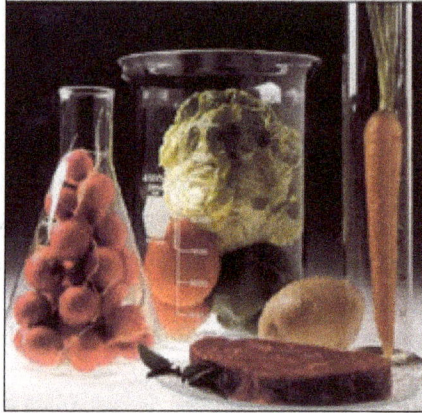

Natural Food Preservatives

In the category of natural food preservatives comes the salt, sugar, alcohol, vinegar etc. These are the traditional preservatives in food that are also used at home while making pickles, jams and juices etc. Also the freezing, boiling, smoking, salting are considered to be the natural ways of preserving food. Coffee powder and soup are dehydrated and freeze-dried for preservation. The citrus food preservatives like citrus acid and ascorbic acid work on enzymes and disrupt their metabolism leading to the preservation.

Sugar and salt are the earliest natural food preservatives that very efficiently drops the growth of bacteria in food. To preserve meat and fish, salt is still used as a natural food preservative.

Chemical Food Preservatives

Chemical food preservatives are also being used for quite some time now. They seem to be the best and the most effective for a longer shelf life and are generally fool proof for the preservation purpose. Examples of chemical food preservatives are:

- Benzoates (such as sodium benzoate, benzoic acid).
- Nitrites (such as sodium nitrite).
- Sulphites (such as sulphur dioxide).
- Sorbates (such as sodium sorbate, potassium sorbate.

Antioxidants are also the chemical food preservatives that act as free radical scavengers. In this category of preservatives in food comes the vitamin C, BHA (butylated hydroxyanisole), bacterial growth inhibitors like sodium nitrite, sulfur dioxide and benzoic acid.

Then there is ethanol that is a one of the chemical preservatives in food, wine and food stored in brandy. Unlike natural food preservatives some of the chemical food preservatives are harmful. Sulfur dioxide and nitrites are the examples. Sulfur dioxide causes irritation in bronchial tubes and nitrites are carcinogenic.

Artificial Preservatives

Artificial preservatives are the chemical substances that stops of delayed the growth of bacteria, spoilage and its discoloration. These artificial preservatives can be added to the food or sprayed on the food.

Types of Artificial Preservatives Food

- Antimicrobial agents,
- Antioxidants,
- Chelating agent.

In antimicrobial comes the Benzoates, Sodium benzoate, Sorbates and Nitrites.

Antioxidants include the Sulfites, Vitamin E, Vitamin C and Butylated hydroxytoluene (BHT)

Chelating agent has the Disodium ethylenediaminetetraacetic acid (EDTA), Polyphosphates and Citric acid

Harmful Food Preservatives

Although preservatives food additives are used to keep the food fresh and to stop the bacterial growth. But still there are certain preservatives in food that are harmful if taken in more than the prescribed limits.

Benzoates

This group of chemical food preservative has been banned in Russia because of its role in triggering allergies, asthma and skin rashes. It is also considered to cause the brain damage. This food preservative is used in fruit juices, tea, coffee etc.

Butylates

This chemical food preservative is expected to cause high blood pressure and cholestrol level. This can affect the kidney and live function. It is found in butter, vegetable oils and margarine.

Butylated Hydroxyanisole

Butylated Hydroxyanisole (BHA) is expected to cause the live diseases and cancer. This food preservative is used to preserve the fresh pork and pork sausages, potato chips, instant teas, cake mixes and many more.

Caramel

Caramel is the coloring agent that causes the vitamin B6 deficiencies, genetic effects and cancer. It is found in candies, bread, brown colored food and frozen pizza.

In addition to this there are many other harmful food preservatives. These are Bromates, Caffeine, Carrageenan, Chlorines, Coal Tar AZO Dies, Gallates, Glutamates, Mono- and Di-glycerides, Nitrates/Nitrites, Saccharin, Sodium Erythrobate, Sulphites and Tannin.

Preservatives Food Additives

All of these chemicals act as either antimicrobials or antioxidants or both. They either inhibit the activity of or kill the bacteria, molds, insects and other microorganisms. Antimicrobials, prevent the growth of molds, yeasts and bacteria and antioxidants keep foods from becoming rancid or developing black spots. They suppress the reaction when foods comes in contact with oxygen, heat, and some metals. They also prevent the loss of some essential amino acids some vitamins.

Table: Some common preservatives and their primary activity.

Chemical Affected	Organism	Action	Use in Foods
Sulfites	Insects and Microorganisms	Antioxidant	Dried Fruits, Wine, Juice
Sodium Nitrite	Clostridia	Antimicrobial	Cured Meats
Propionic Acid	Molds	Antimicrobial	Bread, Cakes, Cheeses
Sorbic Acid	Molds	Antimicrobial	Cheeses, Cakes, Salad Dressing
Benzoic Acid	Yeasts & Molds	Antimicrobial	Soft Drinks, Ketchup, Salad Dressings

There are other antioxidants like Sodium Erythorbate, Erythorbic Acid, Sodium Diacetate, Sodium Succinate, Grape Seed Extract, Pine Bark Extract, Apple Extract Tea Proplyphenols, Succinic Acid and Ascorbic Acid and food preservatives like Parabens and Sodium Dehydro Acetate used frequently for preservation.

TRADITIONAL METHODS OF FOOD PRESERVATION

Food Drying

Food drying is a method of food preservation in which food is dried (dehydrated or desiccated). Drying inhibits the growth of bacteria, yeasts, and mold through the removal of water. Dehydration has been used widely for this purpose since ancient times; the earliest known practice is 12,000 B.C. by inhabitants of the modern Middle

East and Asia regions. Water is traditionally removed through evaporation (air drying, sun drying, smoking or wind drying), although today electric food dehydrators or freeze-drying can be used to speed the drying process and ensure more consistent results.

Flattened fish drying in the sun. Fish are preserved
through such traditional methods as drying, smoking and salting.

A whole potato, sliced pieces (left), and dried sliced pieces (right).

Food Types

Many different foods can be prepared by dehydration. Meat has held a historically significant role. For centuries, much of the European diet depended on dried cod—known as salt cod, bacalhau (with salt), or stockfish (without). Dried fish most commonly cod or haddock, known as Harðfiskur, is a delicacy in Iceland, while dried reindeer meat is a traditional Sami food. Dried meats include prosciutto (Parma ham), bresaola, biltong and beef jerky.

Dried fruits have been consumed historically due to their high sugar content and sweet taste, and a longer shelf-life from drying. Fruits may be used differently when dried. The plum becomes a prune, the grape a raisin. Figs and dates may be transformed into different products that can either be eaten as they are, used in recipes, or rehydrated.

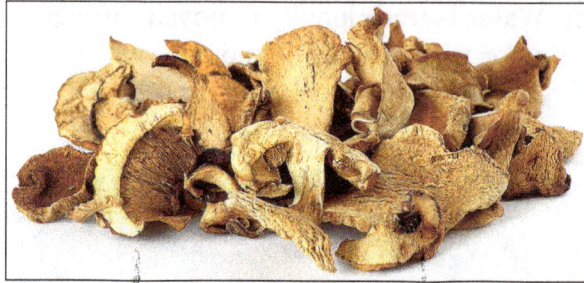
A collection of dried mushrooms.

Sun-drying octopus.

Freeze-dried vegetables are often found in food for backpackers, hunters, and the military. Garlic and onion are often dried. Edible mushrooms, as well as other fungi, are also sometimes dried for preservation purposes or to be used as seasonings.

Preparation

Home drying of vegetables, fruit and meat can be carried out with electrical dehydrators (household appliance) or by sun-drying or by wind. Preservatives such as potassium metabisulfite, BHA, or BHT may be used, but are not required. However, dried products without these preservatives may require refrigeration or freezing to ensure safe storage for a long time.

Industrial food dehydration is often accomplished by freeze-drying. In this case food is flash frozen and put into a reduced-pressure system which causes the water to sublimate directly from the solid to the gaseous phase. Although freeze-drying is more expensive than traditional dehydration techniques. It also mitigates the change in flavor, texture, and nutritional value. In addition, another widely used industrial method of drying of food is convective hot air drying. Industrial hot air dryers are simple and easy to design, construct and maintain. More so, it is very affordable and has been reported to retain most of the nutritional properties of food if dried using appropriate drying conditions.

Another form of food dehydration is irradiation. Irradiation uses x-rays, ultraviolet light, and ionizing radiations to penetrate food to the point of sterilization. Astronauts and people who are highly at risk for microbial infections benefit from this method of food drying.

Hurdle technology is the combination of multiple food preservation methods. Hurdle technology uses low doses of multiple food preservation techniques in order to ensure food is not only safe but is desirable visually and texturally.

Packaging

Packaging ensures effective food preservation. Some methods of packaging that are beneficial to dehydrated food are vacuum sealed, inert gases, or gases that help regulate respiration, biological organisms, and growth of microorganisms.

Salting of Food

Sea salt being added to raw ham to make Prosciutto.

Curing salt or pink salt. It's typically a combination of salt and sodium nitrite, with the pink color added to distinguish it from ordinary salt.

Salting is the preservation of food with dry edible salt. It is related to pickling in general and more specifically to brining (preparing food with brine, that is, salty water) and is one form of curing. It is one of the oldest methods of preserving food, and two historically significant salt-cured foods are salted fish (usually dried and salted cod or salted herring) and salt-cured meat (such as bacon). Vegetables such as runner beans and cabbage are also often preserved in this manner.

Salting is used because most bacteria, fungi and other potentially pathogenic organisms cannot survive in a highly salty environment, due to the hypertonic nature of salt. Any living cell in such an environment will become dehydrated through osmosis and die or become temporarily inactivated.

It was discovered in the 19th century that salt mixed with nitrates (saltpeter) would color meats red, rather than grey, and consumers at that time then strongly preferred the red-colored meat. The food hence preserved stays healthy and fresh for days avoiding bacterial decay.

Pickling

A jar of pickled cucumbers (front) and a jar of pickled onions (back).

Pickling is the process of preserving or extending the lifespan of food by either anaerobic fermentation in brine or immersion in vinegar. In East Asia, vinaigrette (vegetable oil and vinegar) is also used as a pickling medium. The pickling procedure typically affects the food's texture, taste and flavor. The resulting food is called a *pickle*, or, to prevent ambiguity, prefaced with *pickled*. Foods that are pickled include vegetables, fruits, meats, fish, dairy and eggs.

A distinguishing characteristic is a pH of 4.6 or lower, which is sufficient to kill most bacteria. Pickling can preserve perishable foods for months. Antimicrobial herbs and spices, such as mustard seed, garlic, cinnamon or cloves, are often added. If the food contains sufficient moisture, a pickling brine may be produced simply by adding dry salt. For example, sauerkraut and Korean kimchi are produced by salting the vegetables to draw out excess water. Natural fermentation at room temperature, by lactic acid bacteria, produces the required acidity. Other pickles are made by placing vegetables in vinegar. Like the canning process, pickling (which includes fermentation) does not require that the food be completely sterile before it is sealed. The acidity or salinity of the solution, the temperature of fermentation, and the exclusion of oxygen determine which microorganisms dominate, and determine the flavor of the end product.

When both salt concentration and temperature are low, *Leuconostoc mesenteroides* dominates, producing a mix of acids, alcohol, and aroma compounds. At higher temperatures *Lactobacillus plantarum* dominates, which produces primarily lactic acid. Many pickles start with *Leuconostoc*, and change to *Lactobacillus* with higher acidity.

Process

Bát Tràng porcelain vessel for pickling.

In traditional pickling, fruit or vegetables are submerged in brine (20-40 grams/L of salt (3.2–6.4 oz/imp gal or 2.7–5.3 oz/US gal)), or shredded and salted as in sauerkraut preparation, and held underwater by flat stones layered on top. Alternatively, a lid with an airtrap or a tight lid may be used if the lid is able to release pressure which may result from carbon dioxide buildup. Mold or (white) kahm yeast may form on the surface; kahm yeast is mostly harmless but can impart an off taste and may be removed without affecting the pickling process.

In chemical pickling, the fruits or vegetables to be pickled are placed in a sterilized jar along with brine, vinegar, or both, as well as spices, and are then allowed to mature until the desired taste is obtained.

The food can be pre-soaked in brine before transferring to vinegar. This reduces the water content of the food, which would otherwise dilute the vinegar. This method is particularly useful for fruit and vegetables with a high natural water content.

In commercial pickling, a preservative such as sodium benzoate or EDTA may also be added to enhance shelf life. In fermentation pickling, the food itself produces the preservation agent, typically by a process involving *Lactobacillus* bacteria that produce lactic acid as the preservative agent.

Alum is used in pickling to promote crisp texture and is approved as a food additive by the United States Food and Drug Administration.

"Refrigerator pickles" are unfermented pickles made by marinating fruit or vegetables in a seasoned vinegar solution. They must be stored under refrigeration or undergo canning to achieve long-term storage.

Japanese Tsukemono use a variety of pickling ingredients depending on their type, and are produced by combining these ingredients with the vegetables to be preserved and putting the mixture under pressure.

Fermentation

Fermentation in progress: Bubbles of CO_2 form
a froth on top of the fermentation mixture.

Fermentation is a metabolic process that produces chemical changes in organic substrates through the action of enzymes. In biochemistry, it is narrowly defined as the extraction of energy from carbohydrates in the absence of oxygen. In the context of food production, it may more broadly refer to any process in which the activity of microorganisms brings about a desirable change to a foodstuff or beverage. The science of fermentation is known as zymology.

In microorganisms, fermentation is the primary means of producing ATP by the degradation of organic nutrients anaerobically. Humans have used fermentation to produce foodstuffs and beverages since the Neolithic age. For example, fermentation is used for preservation in a process that produces lactic acid found in such sour foods as pickled cucumbers, kimchi, and yogurt, as well as for producing alcoholic beverages such as wine and beer. Fermentation also occurs within the gastrointestinal tracts of all animals, including humans.

Below are some definitions of fermentation. They range from informal, general usages to more scientific definitions.

1. Preservation methods for food via microorganisms.

2. Any process that produces alcoholic beverages or acidic dairy products.

3. Any large-scale microbial process occurring with or without air.

4. Any energy-releasing metabolic process that takes place only under anaerobic conditions.

5. Any metabolic process that releases energy from a sugar or other organic molecule, does not require oxygen or an electron transport system, and uses an organic molecule as the final electron acceptor.

Biological Role

Along with photosynthesis and aerobic respiration, fermentation is a way of extracting energy from molecules, but it is the only one common to all bacteria and eukaryotes. It is therefore considered the oldest metabolic pathway, suitable for an environment that does not yet have oxygen. Yeast, a form of fungus, occurs in almost any environment capable of supporting microbes, from the skins of fruits to the guts of insects and mammals and the deep ocean, and they harvest sugar-rich materials to produce ethanol and carbon dioxide.

The basic mechanism for fermentation remains present in all cells of higher organisms. Mammalian muscle carries out the fermentation that occurs during periods of intense exercise where oxygen supply becomes limited, resulting in the creation of lactic acid. In invertebrates, fermentation also produces succinate and alanine.

Fermentative bacteria play an essential role in the production of methane in habitats ranging from the rumens of cattle to sewage digesters and freshwater sediments. They produce hydrogen, carbon dioxide, formate and acetate and carboxylic acids; and then consortia of microbes convert the carbon dioxide and acetate to methane. Acetogenic bacteria oxidize the acids, obtaining more acetate and either hydrogen or formate. Finally, methanogens (which are in the domain *Archea*) convert acetate to methane.

Biochemical

Fermentation reacts NADH with an endogenous, organic electron acceptor. Usually this is pyruvate formed from sugar through glycolysis. The reaction produces NAD+ and an organic product, typical examples being ethanol, lactic acid, carbon dioxide, and hydrogen gas (H_2). However, more exotic compounds can be produced by fermentation, such as butyric acid and acetone. Fermentation products contain chemical energy (they are not fully oxidized), but are considered waste products, since they cannot be metabolized further without the use of oxygen.

Fermentation normally occurs in an anaerobic environment. In the presence of O_2, NADH, and pyruvate are used to generate ATP in respiration. This is called oxidative phosphorylation, and it generates much more ATP than glycolysis alone. For that reason, fermentation is rarely utilized when oxygen is available. However, even in the presence of abundant oxygen, some strains of yeast such as *Saccharomyces cerevisiae* prefer fermentation to aerobic respiration as long as there is an adequate supply of sugars (a phenomenon known as the Crabtree effect). Some fermentation processes involve obligate anaerobes, which cannot tolerate oxygen.

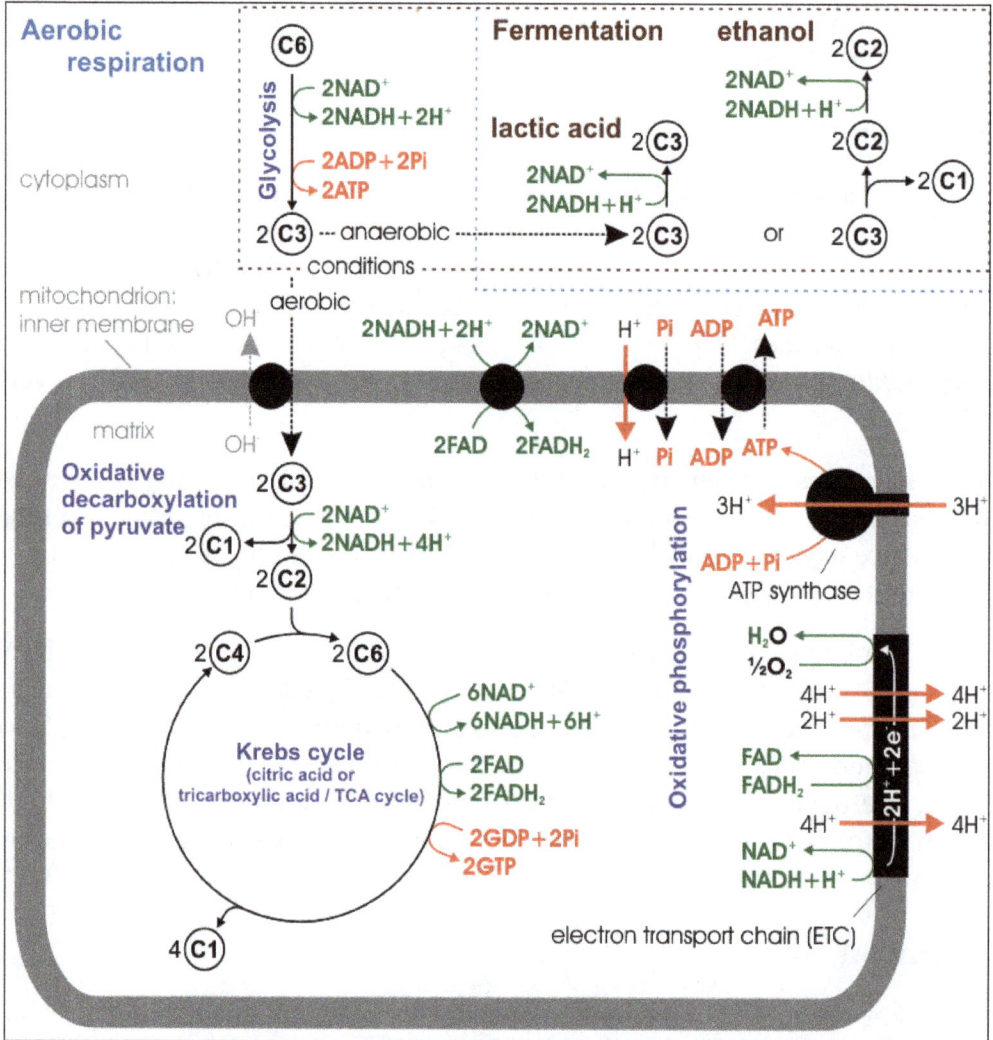

Comparison of aerobic respiration and most known fermentation types in eucaryotic cell. Numbers in circles indicate counts of carbon atoms in molecules, C6 is glucose $C_6H_{12}O_6$, C1 carbon dioxide CO_2. Mitochondrial outer membrane is omitted.

Although yeast carries out the fermentation in the production of ethanol in beers, wines, and other alcoholic drinks, this is not the only possible agent: bacteria carry out the fermentation in the production of xanthan gum.

Products

Ethanol

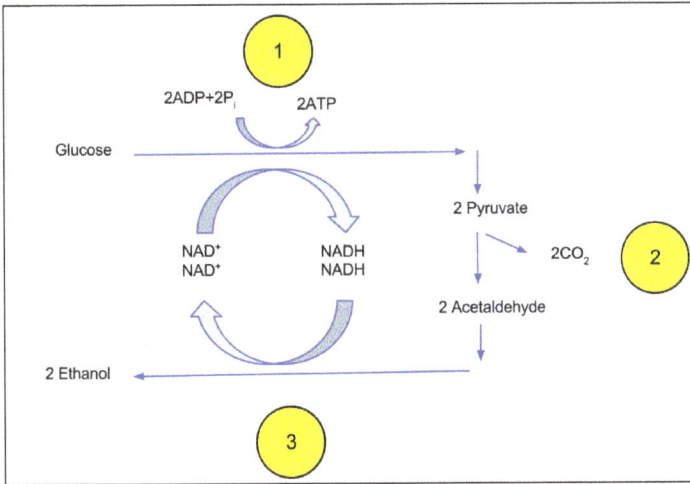

Overview of ethanol fermentation.

In ethanol fermentation, one glucose molecule is converted into two ethanol molecules and two carbon dioxide molecules. It is used to make bread dough rise: the carbon dioxide forms bubbles, expanding the dough into a foam. The ethanol is the intoxicating agent in alcoholic beverages such as wine, beer and liquor. Fermentation of feedstocks, including sugarcane, corn, and sugar beets, produces ethanol that is added to gasoline. In some species of fish, including goldfish and carp, it provides energy when oxygen is scarce (along with lactic acid fermentation).

The figure illustrates the process. Before fermentation, a glucose molecule breaks down into two pyruvate molecules. The energy from this exothermic reaction is used to bind inorganic phosphates to ATP and convert NAD+ to NADH. The pyruvates break down into two acetaldehyde molecules and give off two carbon dioxide molecules as a waste product. The acetaldehyde is reduced into ethanol using the energy and hydrogen from NADH, and the NADH is oxidized into NAD+ so that the cycle may repeat. The reaction is catalysed by the enzymes pyruvate decarboxylase and alcohol dehydrogenase.

Lactic Acid

Homolactic fermentation (producing only lactic acid) is the simplest type of fermentation. The pyruvate from glycolysis undergoes a simple redox reaction, forming lactic acid. It is unique because it is one of the only respiration processes to not produce a gas as a byproduct. Overall, one molecule of glucose (or any six-carbon sugar) is converted to two molecules of lactic acid:

$$C_6H_{12}O_6 \rightarrow 2 \ CH_3CHOHCOOH$$

It occurs in the muscles of animals when they need energy faster than the blood can supply oxygen. It also occurs in some kinds of bacteria (such as lactobacilli) and some fungi. It is the type of bacteria that converts lactose into lactic acid in yogurt, giving it its sour taste. These lactic acid bacteria can carry out either homolactic fermentation, where the end-product is mostly lactic acid, or *Heterolactic fermentation*, where some lactate is further metabolized and results in ethanol and carbon dioxide (via the phosphoketolase pathway), acetate, or other metabolic products, e.g.:

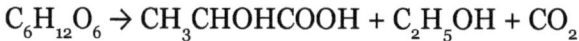

$$C_6H_{12}O_6 \rightarrow CH_3CHOHCOOH + C_2H_5OH + CO_2$$

If lactose is fermented (as in yogurts and cheeses), it is first converted into glucose and galactose (both six-carbon sugars with the same atomic formula):

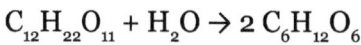

$$C_{12}H_{22}O_{11} + H_2O \rightarrow 2\ C_6H_{12}O_6$$

Heterolactic fermentation is in a sense intermediate between lactic acid fermentation and other types, e.g. alcoholic fermentation. The reasons to go further and convert lactic acid into anything else are:

- The acidity of lactic acid impedes biological processes. This can be beneficial to the fermenting organism as it drives out competitors that are unadapted to the acidity. As a result, the food will have a longer shelf life (part of the reason foods are purposely fermented in the first place); however, beyond a certain point, the acidity starts affecting the organism that produces it.

- The high concentration of lactic acid (the final product of fermentation) drives the equilibrium backwards (Le Chatelier's principle), decreasing the rate at which fermentation can occur and slowing down growth.

- Ethanol, into which lactic acid can be easily converted, is volatile and will readily escape, allowing the reaction to proceed easily. CO_2 is also produced, but it is only weakly acidic and even more volatile than ethanol.

- Acetic acid (another conversion product) is acidic and not as volatile as ethanol; however, in the presence of limited oxygen, its creation from lactic acid releases additional energy. It is a lighter molecule than lactic acid, that forms fewer hydrogen bonds with its surroundings (due to having fewer groups that can form such bonds), thus is more volatile and will also allow the reaction to move forward more quickly.

- If propionic acid, butyric acid, and longer monocarboxylic acids are produced, the amount of acidity produced per glucose consumed will decrease, as with ethanol, allowing faster growth.

Hydrogen Gas

Hydrogen gas is produced in many types of fermentation (mixed acid fermentation, butyric acid fermentation, caproate fermentation, butanol fermentation, glyoxylate fermentation) as a way to regenerate NAD^+ from NADH. Electrons are transferred to ferredoxin, which in turn is oxidized by hydrogenase, producing H_2. Hydrogen gas is a substrate for methanogens and sulfate reducers, which keep the concentration of hydrogen low and favor the production of such an energy-rich compound, but hydrogen gas at a fairly high concentration can nevertheless be formed, as in flatus.

As an example of mixed acid fermentation, bacteria such as *Clostridium pasteurianum* ferment glucose producing butyrate, acetate, carbon dioxide, and hydrogen gas: The reaction leading to acetate is:

$$C_6H_{12}O_6 + 4\,H_2O \rightarrow 2\,CH_3COO^- + 2\,HCO_3^- + 4\,H^+ + 4\,H_2$$

Glucose could theoretically be converted into just CO_2 and H_2, but the global reaction releases little energy.

Modes of Operation

Most industrial fermentation uses batch or fed-batch procedures, although continuous fermentation can be more economical if various challenges, particularly the difficulty of maintaining sterility, can be met.

Batch

In a batch process, all the ingredients are combined and the reactions proceed without any further input. Batch fermentation has been used for millennia to make bread and alcoholic beverages, and it is still a common method, especially when the process is not well understood. However, it can be expensive because the fermentor must be sterilized using high pressure steam between batches. Strictly speaking, there is often addition of small quantities of chemicals to control the pH or suppress foaming.

Batch fermentation goes through a series of phases. There is a lag phase in which cells adjust to their environment; then a phase in which exponential growth occurs. Once many of the nutrients have been consumed, the growth slows and becomes non-exponential, but production of *secondary metabolites* (including commercially important antibiotics and enzymes) accelerates. This continues through a stationary phase after most of the nutrients have been consumed, and then the cells die.

Fed-batch

Fed-batch fermentation is a variation of batch fermentation where some of the ingredients are added during the fermentation. This allows greater control over the stages of the process. In particular, production of secondary metabolites can be increased by adding a limited quantity of nutrients during the non-exponential growth phase. Fed-batch operations are often sandwiched between batch operations.

Open

The high cost of sterilizing the fermentor between batches can be avoided using various open fermentation approaches that are able to resist contamination. One is to use a naturally evolved mixed culture. This is particularly favored in wastewater treatment, since mixed populations can adapt to a wide variety of wastes. Thermophilic bacteria can produce lactic acid at temperatures of around 50 degrees Celsius, sufficient to discourage microbial contamination; and ethanol has been produced at a temperature of 70 °C. This is just below its boiling point (78 °C), making it easy to extract. Halophilic bacteria can produce bioplastics in hypersaline conditions. Solid-state fermentation adds a small amount of water to a solid substrate; it is widely used in the food industry to produce flavors, enzymes and organic acids.

Continuous

In continuous fermentation, substrates are added and final products removed continuously. There are three varieties: chemostats, which hold nutrient levels constant; turbidostats, which keep cell mass constant; and plug flow reactors in which the culture medium flows steadily through a tube while the cells are recycled from the outlet to the inlet. If the process works well, there is a steady flow of feed and effluent and the costs of repeatedly setting up a batch are avoided. Also, it can prolong the exponential growth phase and avoid byproducts that inhibit the reactions by continuously removing them. However, it is difficult to maintain a steady state and avoid contamination, and the design tends to be complex. Typically the fermentor must run for over 500 hours to be more economical than batch processors.

THERMAL PROCESSING OF FOOD

Thermal processing is a commercial technique used to sterilize food through the use of high temperatures. The primary purpose of thermal processing is to destroy potential toxins in food. The process does have limitations and its application must be carefully overseen by an authority who understands the importance of variables in regulating thermal processing.

Thermal processing is a food sterilization technique in which the food is heated at a temperature high enough to destroy microbes and enzymes. The specific amount of time required depends upon the specific food and the growth habits of the enzymes or microbes. Both the texture and the nutritional content of the food may be altered due to thermal processing.

Methods

Food may be sterilized using in-package sterilization techniques. Using this technique, the food is sterilized while it is already in a bottle, can or other package. The other option is UHT (ultra-high temperature) or aseptically processed products, which require the package and the food to be sterilized using thermal processing separately before they are sealed together.

Acids

The presence of acids alters the temperature and processing times required for thermal processing. Some types of spores cannot reproduce in acidic environments, while acid helps aid in the destruction of other microorganisms.

Limitations

In order to be classified as having commercial sterility, all the microorganisms do not have to be destroyed. Commercial sterility implies only that any remaining microbes will be incapable of continuing to grow and thrive in the food.

Blanching

The primary purpose of blanching is to destroy enzyme activity in fruit and vegetables. It is not intended as a sole method of preservation, but as a pretreatment prior to freezing, drying and canning. Other functions of blanching include:

- Reducing surface microbial contamination.

- Softening vegetable tissues to facilitate filling into containers.

- Removing air from intercellular spaces prior to canning.

Blanching and Enzyme Inactivation

Freezing and dehydration are insuffcient to inactivate enzymes and therefore blanching can be employed. Canning conditions may allow suffcient time for enzyme activity. Enzymes are proteins which are denatured at high temperatures and lose their activity. Enzymes which cause loss of quality include Lipoxygenase, Polyphenoloxidase, Polygaacturonase and Chlorophyllase. Heat resistant enzymes include Catalase and Peroxidase

Methods of Blanching

Blanching is carried out at up to 100 °C using hot water or steam at or near atmospheric pressure.

Some use of fluidised bed blanchers, utilising a mixture of air and steam, has been reported. Advantages include faster, more uniform heating, good mixing of the product, reduction in effluent, shorter processing time and hence reduced loss of soluble and heat sensitive components.

There is also some use of microwaves for blanching. Advantages include rapid heating and less loss of water soluble components. Disadvantages include high capital costs and potential diffculties in uniformity of heating.

Steam Blanchers

This is the preferred method for foods with large cut surface areas as lower leaching losses. Normally food material carried on a mesh belt or rotatory cylinder through a steam atmosphere, residence time controlled by speed of the conveyor or rotation. Often poor uniformity of heating in the multiple layers of food, so attaining the required time-temperature at the centre results in overheating of outside layers.

Individual Quick Blanching (IQB) involves a first stage in which a single layer of the food is heated to suffcient temperature to inactivate enzymes and a second stage in which a deep bed of the product is held for suffcient time to allow the temperature at the centre of each piece to increase to that needed for inactivation.

The reduced heating time (e.g. for 10 mm diced carrot, 25 s heating and 50 s holding compared with 3 minutes conventional blanching) results in higher energy effcienciess For small products (e.g. peas, sliced or diced carrots), mass of produce blanched per kg steam increases from 0.5 kg for conventional steam blanchers to 6-7 kg for IQB.

Hot Water Blanchers

Includes various designs which hold the food in hot water (70 to 100 °C) for a specified time, then moves it to a dewatering/cooling section. In blanchers of this type the food enters a slowly rotating drum, partially submerged in the hot water. It is carried along by internal flights, residence time being controlled by the speed of rotation.

Pipe blanchers consist of insulated tubes through which hot water is circulated. Food is metered into the stream, residence time being controlled by the length of the pipe and velocity of the water.

The blancher-cooker has three sections, a preheating stage, a blanching stage, and a cooling stage. As the food remains on a single belt throughout the process, it is less likely to be physically damaged. With the heat recovery incorporated in the system, 16 to 20 kg of product can be blanched for every kg of steam, compared with 0.25 to 0.5 per kg stream in the conventional hot water blanchers.

Testing of the Effectiveness of Blanching

Over blanching causes quality loss due to overheating while under blanching causes quality loss due to increased enzyme activity because enzymes activated and substrates released by heat. The Peroxidase test in vegetables is used to detect enzyme inactivation. This enzyme is not in itself implicated in degradation, but is relatively heat resistant and easily detected. It consists of adding guaiacol solution and hydrogen peroxide solution and observing the development of a brown colour indicating peroxidase activity.

Complete inactivation is not always essential – green beans, peas and carrots with some residual peroxidase activity have shown adequate storage quality at -20 °C through with other vegetable (e.g. Brussels sprouts) zero peroxidase activity is essential.

Sterilisation

Unlike pasteurized products where the survival of heat resistant microorganisms is accepted, the aim of sterilization is the destruction of all bacteria including their spores. Heat treatment of such products must be severe enough to inactivate/kill the most heat resistant bacterial microorganisms, which are the spores of Bacillus and Clostridium. Food products filled in sealed containers are exposed to temperatures above 100 °C in pressure cookers. Temperatures above 100 °C, usually ranging from 110-121 °C depending on the type of product, must be reached inside the product. Products are kept for a defined period of time at temperature levels required for the sterilization depending on type of product and size of container.

If spores are not completely inactivated, vegetative microorganisms will grow from the spores as soon as conditions are favourable again. Favourable conditions will exist when the heat treatment is completed and the products are stored under ambient temperatures. The surviving microorganisms can either spoil preserved food or produce toxins which cause food poisoning. Amongst the two groups of spore producing microorganisms Clostridium is more heat resistant than Bacillus. Temperatures of 110 °C will kill most Bacillus spores within a short time. In the case of Clostridium temperatures of up to 121 °C are needed to kill the spores within a relatively short time. These sterilization temperatures are needed for short-term inactivation (within a few seconds) of spores of Bacillus or Clostridium. These spores can also be killed at slightly lower temperatures, but longer heat treatment periods must be applied.

From the microbial point of view, it would be ideal to employ very intensive heat treatment which would eliminate the risk of any surviving microorganisms. However, most food products cannot be submitted to such intensive heat stress without suffering degradation of their sensory quality or loss of nutritional value (destruction of vitamins and protein components). In order to comply with above aspects, a compromise has to be reached in order to keep the heat sterilization intensive enough for the microbiological safety of the products and as moderate as possible for product quality reasons.

"Commercial sterility" implies less than absolute destruction of all micro-organisms and spores, but any remaining would be incapable of growth in the food under existing conditions. Time-temperature combination required to inactivate most heat resistant pathogens and spoilage organisms. Most heat resistant pathogen is Clostridium botulinum. Most heat-resistant (non-pathogenic) spoilage microorganisms are Bacillus stearothermophilis and Clostridium thermosaccharolytom. Severity of treatment can result in substantial changes to nutritive and sensory characteristics. Two typical forms of sterilised product are:

- In package sterilised, in which product is packed into containers and the container of product is then sterilised, e.g. canning, some bottled products, retort pouches.

- UHT or Aseptically processed products in which the product and the package is sterilised separately then the package is filled with the sterile product and sealed under specific conditions, e.g. long life milk, tetrapack or combibloc fruit juices and soups etc.

Canned Foods

Canned foods are processed so that they are shelf stable. They should be 'commercially sterile'. That means if any microbes survive the processing, they should not be capable of growing (and therefore spoiling the contents) under the normal storage conditions of the can. Most canned foods are sterile (i.e. there are no living organisms present) but some may contain viable organisms which cannot grow because of unsuitable conditions.

- Water,

- Temperature,

- pH,

- Water activity,

- Preservatives.

If a canned food is spoilt by microbial spoilage, examination of the microbial types that caused it can pinpoint the offending errors in processing or handling.

Conditions affecting the Growth of Microorganisms

Water

Water content and the availability of water Aw can affect the growth of microbes in food.

Temperature

Temperature influences the rate of growth of microbes as well as determining which microbes will grow. Microbes grow fastest at their optimum temperature. For convenience microbes can be divided into groups which have similar optimum temperature for growth.

Table: Growth temperatures (°C) for microbial growth.

Group	Min.	Opt.	Max.
Thermophiles	40	55	75
Mesophiles	5	37	45
Psychotrophs	-3	20	30

Oxygen Requirements

Micro-organisms can be classified into three general groups regarding their oxygen requirements.

- Aerobes – Can only grow in the presence of oxygen.

- Anaerobes – Can only grow in the absence of oxygen.

- Facultative Anaerobes – Adaptable, grows best aerobically but can grow anaerobically.

pH

In regard to pH, microbes have ideal pH ranges within which they grow as follows:

Table: pH ranges for microbial growth.

Group	pH
Low acid	>5.0
Medium acid	4.5 - 5.0
Acid	3.7 - 4.5
High acid	< 3.7

Types of Microorganisms Important in Retorted Foods

A number of organisms are important when it comes to the safe processing of canned foods.

Table: Microorganisms important for retorted foods.

Type	Species	Description
Thermophilic Spore Formers	Flat Sours - B. sterothermophilus	High heat resistance, product acid, don't produce gas, found in sugar, salt and spices.
	Thermophilic Anaerobes – C. thermosaccharolyticum	High heat resistance, product acid and gas (CO_2).
	Sulphide types – Desutfomotomaculum nigrificans	High heat resistance, produce H_2S.

Mesophilic Spore Formers (The process should be designed to kill these microbes)	C. sporogenes, C. botulinium	Produce gas (CO_2 and sometimes H_2, moderate heat resistance.
	Bacillus spp – B. polymyxa, B. macerans etc	Moderate to low heat resistance, some may grow in acid foods.
Non Spore Forming Microbes	Various	Occur only in grossly under processed or leaking caps.
		Can be almost any microbe depending on acidity of the product.
		May or may not produce gas.
		Usually in mixed populations.

Microbial Spoilage of Canned Foods

There are a number of important factors which can cause spoilage of canned foods.

Table: Factors affecting spoilage of canned foods.

Type	Description
Pre-process spoilage	Delays between filling and retorting can allow microbes to grow and produce gas or spoil food. Retorting kills microbes but the can will be swollen and food spoilage.
Not processed	Filled cans missing retor.
Under processed	Caused by: • Incorrect calculations. • Faulty retort operation. • Operator error e.g. inadequate venting. • Poor retort design e.g. cold spots. • Higher spore load – poor or different raw ingredients. Under processing usually still kills vegetative cells. Survivors are usually mesophilic spore formers or moderate heat resistance.
Thermophilic Spoilage	Canning operations are sometimes not designed to kill thermophiles of high heat resistance (e.g. B.stearothermophilus of D 121.1 = 5 min) as they do not grow below 40 °C. If they survive they will grow if there is either slow cooling or storage at high temperatures. Thermophilic spore formers will be found in pure cultures.

Leaker Spoilage	If can seams are inadequately formed, microbes may enter can after processing, particularly when the can is moist e.g. during cooling. Usual contamination is a mixed of a variety of non-heat resistant microbes.
	Cans may leak food or if leakage point is blocked with food, they can swell.

Sterilisation Process and Equipment

The sterilization process in the canned product can be subdivided into three phases. By means of a heating medium (water or steam) the product temperature is increased from ambient to the required sterilization temperature (phase 1 = heating phase). This temperature is maintained for a defined time (phase 2 = holding phase). In (phase 3 = cooling phase) the temperature in the can is decreased by introduction of cold water into the autoclave.

Autoclaves

In order to reach temperatures above 100 °C ("sterilization"), the thermal treatment has to be performed under pressure in pressure cookers, also called autoclaves or retorts.

In autoclaves or retorts, high temperatures are generated either by direct steam injection, by heating water up to temperatures over 100 °C or by combined steam and water heating. The autoclave must be fitted with a thermometer, pressure gauge, pressure relief valve, vent to manually release pressure, safety relief valve where steam is released when reaching a certain pressure, water supply valve and a steam supply valve. The steam supply valve is applicable when the autoclave is run with steam as the sterilization medium or when steam is used for heating up the sterilization medium water.

Simple Small Autoclaves

These are usually vertical autoclaves with the lid on top. Through the opened lid the goods to be sterilized are loaded into the autoclave. The cans are normally placed in metal baskets. The baskets are placed in the autoclave, either singly or several stapled on top of each other. Before starting the sterilization, the lid must be firmly locked onto the body of the autoclave. The autoclave and lid are designed to withstand pressures up to 5.0 bar. These types of autoclaves are best suited for smaller operations as they do not require complicated supply lines and should be available at affordable prices.

Larger Autoclaves

These are usually horizontal and loaded through a front lid. Horizontal autoclaves can be built as single or double vessel system. The double vessel systems have the advantage that the water is heated up in the upper vessel to the sterilization temperature and released into the lower (processing) vessel, when it is loaded and hermetically closed. Using the two–vessel system, the heat treatment can begin immediately without lengthy heating up of the processing vessel and the hot water can be recycled afterwards for immediate use in the following sterilization cycle.

If steam is used instead of water as the sterilization medium, the injection of steam into a single vessel autoclave will instantly build up the autoclave temperature desired for the process.

Rotary Autoclaves

Another technology employed is rotary autoclaves in which the basket containing the cans rotates during sterilization. This technique is useful for cans with liquid or semi-liquid content as it achieves a mixing effect of the liquid/semi-liquid goods resulting in accelerated heat penetration. The sterilization process can be kept shorter and better sensory quality of the goods is ensured.

At the final stage of the sterilization process the products must be cooled down as quickly as possible. This operation is done in the autoclave by introducing cold water. The contact of cold water with steam causes the latter to condense with a rapid pressure drop in the retort. However, the overpressure built up during thermal treatment within the cans, jars or pouches remains for a certain period.

During this phase, when the outside pressure is low but the pressure inside the containers is still high due to high temperatures there, the pressure difference may induce permanent deformation of the containers.

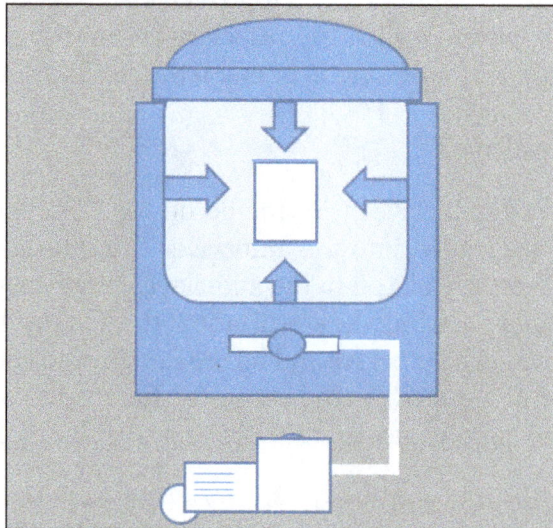

Pressure inside autoclave (blue) and inside cans (red) during heating and cooling phase.

Therefore, high pressure difference between the autoclave and the thermal pressure in the containers must be avoided. This is generally achieved by a blast of compressed air into the autoclave at the initial phase of the coolings. Suffcient hydrostatic pressure of the introduced cooling water can also build up counter pressure so that in specific cases, in particular where strong resistant metallic cans are used, the water pressure can be suffcie and compressed air may not be needed. For the stabilization of metallic cans, stabilization rims can be moulded in lids, bottom and bodies.

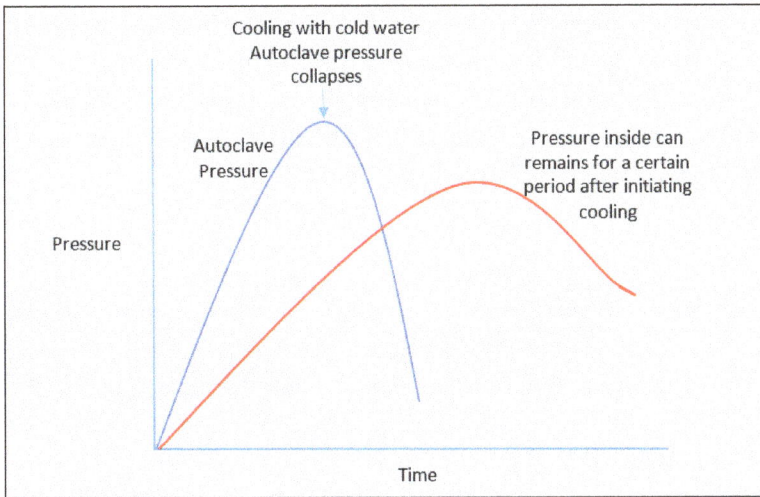

Producing counter pressure on cans inside the autoclave with compressed air.

Types of Containers for Thermally Treated Products

Containers for heat-preserved food must be hermetically sealed and airtight to avoid recontamination from environmental microflora. Most of the thermally preserved products are in metal containers (cans). Others are packed in glass jars or plastic or aluminium/plastic laminated pouches.

Metal Containers are Cans

Produced from tinplate. They are usually cylindrical. However, other shapes such as rectangular or pear-shaped cans also exist. Tinplate consists of steel plate which is electrolytically coated with tin on both sides. The steel body is usually 0.22 to 0.28 mm in thickness. The tin layer is very thin (from 0.38 to 3.08 μm). In addition, the interior of the cans is lined with a synthetic compound to prevent any chemical reaction of the tinplate with the enclosed food.

Tin cans consist of two or three elements. In the case of three-piece steel cans, they are composed of the body and two ends (bottom and lid). The body is made of a thin steel strip, the smaller ends of which are soldered together to a cylindrical shape. Modern cans are induction-soldered and the soldering area is covered inside with a side-strip coating for protection and coverage of the seam. The use of lead soldered food cans was stopped decades ago. Hence the risk of poisonous lead entering canned food no longer exists.

Two-piece steel cans have a lid similar to the three-piece cans but the bottom and body consist of one piece, which is moulded from a circular flat piece of metal into a cup. These cup-shaped parts may be shallow-drawn (with short side wall) or deep-drawn (with longer side walls). However, the length of the side walls is limited through the low moulding ability of steel (example: tuna tins 42/85 mm, i.e. side wall: diameter = 1:2).

Aluminium is frequently used for smaller and easy-to-open cans. Aluminium cans are usually deep-drawn two-piece cans, i.e. the body and the bottom end are formed out of one piece and only the top end is seamed on after the filling operation. The advantages of aluminium cans compared to tin cans are their better deep-drawing capability, low weight, resistance to corrosion, good thermal conductivity and easy recyclability. They are less rigid but more expensive than steel plate cans.

Glass Jars

Glass jars are sometimes used for meat products but are not common due to their fragility. They consist of a glass body and a metal lid. The seaming panel of the metal lid has a lining of synthetic material. Glass lids on jars are fitted by means of a rubber ring.

Retortable Pouches

Retortable pouches, which are containers made either of laminates of synthetic materials only or laminates of aluminium foil with synthetic materials, are of growing importance in thermal food preservation. Thermo-stabilized laminated food pouches, have a seal layer which is usually PP (polypropylene) or PP-PE (polyethylene) polymer, and the outside layers are usually made of polyester (PETP) or nylon. They can be used for frankfurters in brine, ready-to-eat meat dishes etc. From certain laminated films, for instance, polyester/polyethylene (PETP/PE) or polyamide/polyethylene (PA/PE), relatively rigid container can be made, usually by deep drawing.

They are used for pieces of cured ham or other kinds of processed meat. Small can-shaped round containers are made from aluminium foil and polyethylene (PE) or polypropylene (PP) laminate and are widely used for small portions, particularly of sausage mix. PE or PP permits the heat-sealing of the lid made of the same laminate onto these containers, which can then be subjected to intensive heat treatment of 125 °C or above.

One advantage of the retortable pouches/laminated containers is their good thermal conductivity which can considerably reduce the required heat treatment time and hence is beneficial for the sensory product quality.

Cleaning of Containers Prior to Filling

Rigid containers (cans, glass jars) are delivered open to meat processing plants, i.e. with the lids separate. During transport and storage, dust can settle inside the cans, which must be removed prior to filling the cans. This can be done at the small-scale level by manually washing the cans with hot water. Industrial production canning lines are equipped with steam cleaning facilities, where steam is blown into the cans prior to filling.

Seaming of Cans

After the can is filled with the product mix the can is sealed with a tight mechanical structure - the so-called double seam. The double seam, in its final form and shape, consists of three layers of lid (D, black colour) and two layers of body material (D, striated). The layers must overlap significantly and all curves must be of rounded shape to avoid small cracks. Each double seam is achieved in two unit operations referred to as "first operation" (A, B) and "second operation" (C, D).

The can covered with the lid is placed on the base plate of the can seaming machine. The can is moved upwards while the seaming chuck keeps the lid fixed in position. The pressure applied to the can from the base plate can be regulated and must be strong enough to ensure simultaneous movement of the lid and the can to avoid scratching-off of the sealing compound.

In the first operation the lid hook and body hook are interlocked by rolling the two into each other using the seaming roll with the deep and narrow groove. The body hook is now almost parallel to the lid hook and the curl of the lid adjacent to or touching the body wall of the can. In the second operation, the interlocked hooks are pressed together by a seaming roll with a flat and wide groove. Wrinkles are ironed out and the rubber-based material is equally distributed in the seam, filling all existing gaps thus resulting in a hermetically sealed container.

Design of Seaming Rolls

The seaming rolls for the first and second operations are designed differently in order to facilitate the respective operations. The seaming roll for the first operation has a deep but narrow groove to interlock body and lid hock (rolling the hocks into each other). The seaming roll for the second operation has a flat but wide groove to press the interlocked hooks together (sealing the seam). The first action first roll) is rolling (interlocking) the hooks, the second action (second roll) is compressing (sealing) the seam.

Death Rate Curve: D Value

At slightly elevated temperatures most microbes will grow and multiply quickly. At relatively high temperatures, microbes can be destroyed. However, there is a lot of variation within any one population of microbes of the same species – most will be killed relatively quickly, others can survive much longer. If a population of microbes is held at a constant high temperature, the number of surviving spores or cells plotted against time (on a logarithmic scale) will look like the following graph – which is referred to as the 'death rate curve'.

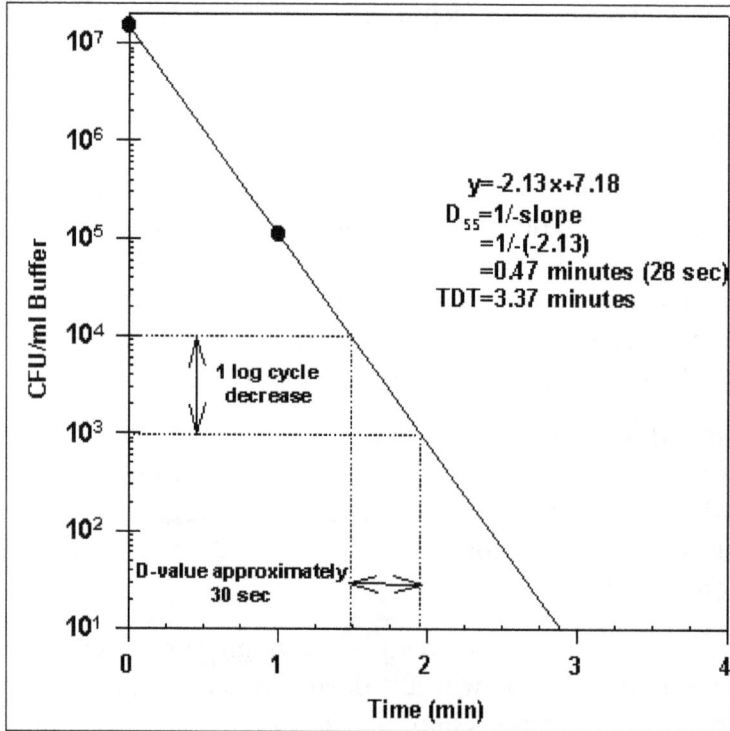

Death Rate Curve (D-value).

This graph is a straight line – it is referred to as the Logarithmic order of death. Logarithms refers to the power to which a base must be raised to produce a given number. For example, if the base is 10, then the logarithm of 1,000 (written log 1,000 or log10 1,000) is 3 because 103 = 1,000. The "death rate curve" is a straight line when plotted using a logarithmic scale – this means that if in some time period the number was reduced from 1000 to 100 (divided by ten, sometimes referred to as "1 log reduction"), then if you had held the microbes at the same temperature for twice that time period, the number would have been reduced to 1 (divided by 100, or "2 log reductions").

The time period for each "log reduction" is referred to as the decimal reduction time or D value. For example the D-value of Bacillus stearothermophillus a common spoilage microorganism at 121 °C is about 4 minutes. This means if you had cans of food product each containing 1000 of these spore and you held the product at a constant temperature of 121 °C:

- After 4 minutes (1 D-value) there would be 100 spores surviving in each can (1 log reduction).

- After 8 minutes (twice D-value) there would be 10 spores surviving in each can (2 log reductions).

- After 12 minutes (3 times D-value) there would be 1 spore surviving in each can (3 log reductions).

If this food product, with an initial count 1000 spores of Bacillus stearothermophillus, was held for 16 minutes at 121 °C it would result in 4 log reductions, or 0.1 spores surviving in each can. 0.1 spores per can means that on average there would be one spore surviving in each group of ten cans. After holding for 20 minutes there would be one spore per 100 cans and so on.

Based on this:

- The higher the number of microbes initially present the longer it takes to reduce the numbers to an acceptable level. Therefore, good quality raw materials and hygienic pre-processing is essential if the commercial sterility of the processed product is to be assured.

- It is theoretically impossible to destroy all cells – therefore we reduce the probability of spoilage to an acceptable small number – perhaps 1 in 1 million. The probability of a pathogen surviving must be even less – perhaps on in one billion or less.

- The above refers to holding the product at a constant temperature. Remember destruction of microbes is temperature dependent – they get killed more quickly at the higher temperatures. Therefore you would expect that if you increase the temperature, decimal reduction time D-value) would decrease.

Thermal Death Time Curve

If D-value versus time is plotted – again on a logarithmic scale, the graph looks very similar to the one previously. This one is called the Thermal death time (TDT) curve. This time the straight line graph means that if you change the temperature by a certain amount, the D-value will change by a factor of 10. If you had changed it by twice that amount, D-value will change by a factor of 100. The change in temperature to cause a factor of the ten change in D-value is referred to as that z-value.

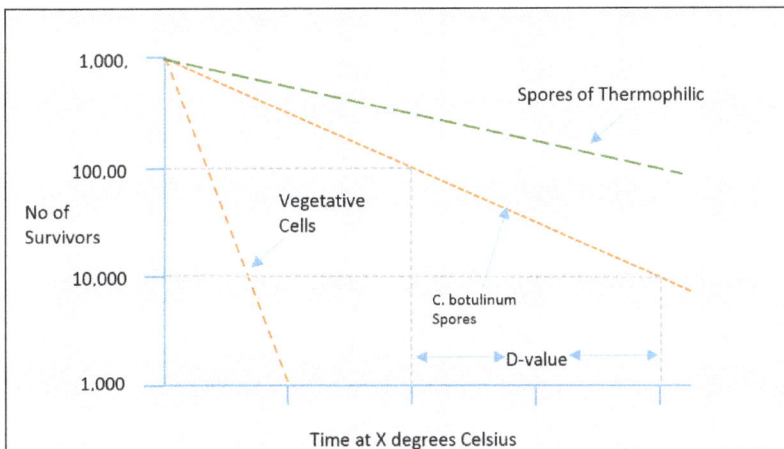

Thermal Death Rate Curves.

The Z-value for Bacillus stearothermophillus is 10 °C. Remember the D-value for this microorganism at 121 °C is 4 minutes. Therefore if you held the containing this microbe at 111 oC (10 oC, or one z-value, less than 121 oC), D-value would be 400 minutes.

That is, for Bacillus stearothermophillus, 4 minutes at 121 °C will have the same effect (one log reduction in spores) as 40 minutes at 111oC, which would have the same effect as 400 minutes at 101 °C. It is obvious why using high processing temperatures is an advantage. The D-values of different microbes differ greatly – for example, the D-value of Clostridium botulinum at 121 °C is about 0.21 minutes. However the z-value of microorganisms is close to 10 °C.

Factors that affect the Heat Resistance of Micro-organisms

A range of factors affect the heat resistance of micro-organisms. The most important are:

- Type of micro-organism – Species and strains differ, spores are more resistant than vegetative cells.

- Conditions during cell growth or spore formation – Spores produced at higher temperature are more heat resistant, stage of growth and the type of medium in which they grow can also affect heat resistance.

- Conditions during heat treatment including pH – Pathogenic and spoilage bacteria are less heat resistant at more acid (low) pH, yeasts and fungi are more acid tolerant but less heat resistant than bacterial spores.

- Aw – Moist heat is more effective than dry heat.

- Composition – Protein, fats and high concentration of sucrose increases heat resistance.

D and z-values of enzymes are generally in a similar range to those of micro-organisms, but some are very heat resistant.

Design of Heat Sterilization Processes

The design of heat processes must:

- Take account of the type of microorganism (determined largely by food conductions e.g. acidity) and its heat resistance.

- Result in an acceptably low probability of survival of spores.

- Be effective in every part of the food.

In low acid foods (pH<4,5), Clostridium botulinum, is the most dangerous, heat resistant spore forming pathogen (D121=0.1 to 0.2 min). It is anaerobic and so can survive and grow in a sealed can. Its destruction is a minimum requirement of heat sterilisation.

This is often interpreted as "12D" process – that is, the product must be treated for 12 times the D-value of the microbe. For Clostridium botulinum this is a process equivalent to about 2.5 minutes at 121 °C – this is commonly known as a "botulinum cook". Normally a more severe heat treatment is required to destroy other more heat resistant spoilage bacterial. For example Bacillus stearothermophillus (thermophile – won't grow at less than 35 °C, so proper can cooling is important) can produce the "flat sour" defect. Its D-value at 121 °C is commonly around 4 min, but it is not of health significance.

In high acid foods (pH < 4.5) the anaerobic pathogens cannot grow or produce toxins. Spoilage microorganisms are quickly killed at temperatures of about 90 °C. Therefore the minimum treatment applied to high acid foods often involves ensuring every part of the product reaches a temperature of at least 95 °C, e.g. pasteurisation. In acid foods where the pH is close to 4.5 (e.g. foods such as tomatoes and pears) Clostridium butyricum can cause spoilage. It is a common soil borne micro-organism, and grows easily on surfaces in the food plant. It is not killed by processes commonly used for acid foods and can cause swelling/bursting of the cans in about 2 weeks.

F_0 Value

The amount of heat treatment applied to a food product can be measured using the F-value-concept. This concept is practiced in canning plants, in particular as part of the HACCP-system. The size and format of cans is of utmost importance for the speed of heat penetration. Temperatures to be achieved at the "cold point" of the can where the heat arrives last, are reached faster in small cans due to the shorter distance to the heat source than in large cans.

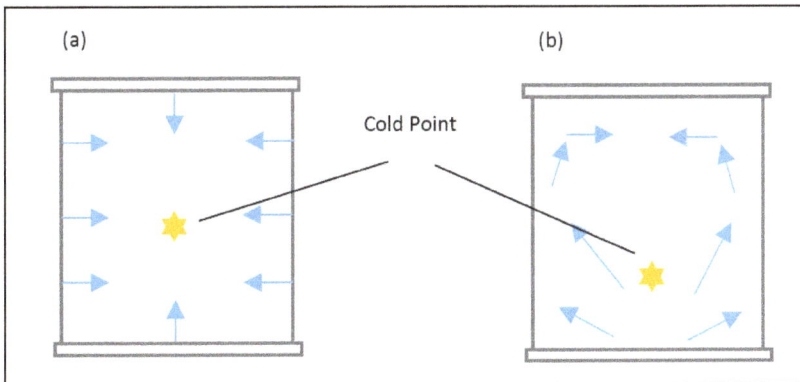

The F_0 value is a measure of the "sterilising value" of a process. It can be thought of as the time required at a temperature of 121 °C to reduce microbial numbers by the same amount as the actual process being considered.

Remember processes are not always carried out at 121 °C and certainly product temperature is not constant at this temperature throughout the process.

It therefore provides a basis for comparing different heat sterilisation procedures if two processes have the same The F_0 value, they provide the same level of sterilisation. It therefore provides a basis for comparing different heat sterilisation procedures if two processes have the same The F_0 value, they provide the same level of sterilisation.

The temperature of 121 °C is simply an arbitrary reference – there is nothing special about this particular temperature. Why choose and off temperature like 121 °C? In the past someone decided 250°F which is equal to 121 °C was a good reference temperature. More accurately it is 121.1 °C.

A similar concept to F_0 often used in determining the heat treatment of beers and other high acid foods is "pasteurising units" (or PU's) – 1 PU is equivalent to pasteurising at 60 °C for one minute. The minimum treatment for low acid products, the "botulinum cook", therefore has a F_0 of 2.5 minutes (i.e. 12 × 0.21 = 2.5 min).

The required level of heat treatment (F_0 of the process) may vary with factors such as pH and carbohydrate level, and type and expected level of contamination with micro-organisms. Other chemical additives may also assist inhibition of microorganisms e.g. salt, alcohol, nitrite and misin (these last two are both 'sporostatic' and stop spores germinating and so enable the use of lesser processing conditions). Also some products require additional processing to achieve the required level of cook e.g. baked beans must be soft enough.

Table: F-values (per minute) for the temperature range of 100 °C to 135 °C.

°C	F-value	°C	F-value
100	0.0077	118	0.4885
101	0.0097	119	0.6150
102	0.0123	120	0.7746
103	0.0154	121	1
104	0.0194	122	1.2270
105	0.0245	123	1.5446
106	0.0308	124	1.9444
107	0.0489	125	2.4480
108	0.0615	126	3.0817
109	0.0775	127	3.8805
110	0.0975	128	4.8852
111	0.1227	129	6.1501
112	0.1545	130	7.459
113	0.1545	131	7.7466
114	0.1945	132	12.2699
115	0.2449	133	15.4560

The Lethality Factor "L" Given that the F_0 is based on a constant reference temperature

ture of 121 °C, but the product is mostly at a different temperature, how can the F_0 be calculated? This is the purpose of the Lethality Factor or "L-value". It is defined as the time at 121.1 °C which is equivalent in sterilising value to one minute at some other temperature. One minute at some temperature will contribute "L" minutes worth of Fo, where "L" is the L_0 value for the temperature concerned. The L-value is dependent on the z-value of the microorganism being considered, but for most purposes z=10 °C. L-value can be calculated from the formula or can be read from a table.

$$L \ = \ 10(T-121.1)/z$$

A product is held at a temperature of 118 °C for a period of 12 minutes. Ignoring other heating and cooling periods, what is the F_0 value of this process? From the formula, the L-value for 118 °C is 0.490. That is each minute at 118 °C contribures 0.490 minutes to the F_0 value. Therefore the Fo value of this process = 12 x 0.490 = 5.9 minutes.

Calculating the F_0 value when temperatures vary: In a real retort process the temperature of the product is not constant – it slowly heats up, will stay at a constant temperature for some time, then cool down again. The period when the product is heating and cooling contribute significantly to the severity of the process. To calculate the F_0 value of such a process, the contribution of the varying temperatures must be converted to an equivalent F_0 value. This is achieved based on the L-value.

Graphical Method

This involves drawing a graph of the product temperature vs time, then looking up the L-value of each temperature, and plotting L-value against time. The area under this graph is a measure of the L-value.

Trapezoidal Integration

For this method, determine the L-value for each temperature measurement, add the L-value together then multiply by the time interval in minutes between temperature measurements (if temperatures are measured every minute there is no need to multiply). Obviously as the severity of the process is related to the time spent at high temperatures the faster a product is heated the greater will be the severity of the process (for the same process time).

A number of factors affect the rate at which a product heats inside a container:

- Type of container – For example glass is not a good conductor of heat so you would expect product in a glass jar to heat more slowly than an equivalent size/shape metal can.

- Size and shape of the container – A large container will take longer to heat than a small container.

- Retort temperature – A higher retort temperate will result in more rapid heating but also may lead to more over processing of product near the package surface.

- Agitation of the containers will increase the heating rate by mixing the contents of the container, especially with viscous or semi-solid foods. End over end agitation is better than axial agitation.

- Type of product – Different products conduct heat more or less easily and have different heat capacities. Some products are more viscous than others which can have a particularly significant effect in agitating retorts. Therefore different products will heat at a different rate.

- Headspace – Insufficient headspace can also affect the rate of heating, especially in an agitating retort.

Therefore if any of these factors change, the severity of the process needs to be reevaluated.

FOOD IRRADIATION

Food irradiation involves the use of either high-speed electron beams or high-energy radiation with wavelengths smaller than 200 nanometres, or 2000 angstroms (e.g., X-rays and gamma rays). These rays contain sufficient energy to break chemical bonds and ionize molecules that lie in their path. The two most common sources of high-energy radiation used in the food industry are cobalt-60 (^{60}Co) and cesium-137 (^{137}Cs). For the same level of energy, gamma rays have a greater penetrating power into foods than high-speed electrons.

The unit of absorbed dose of radiation by a material is denoted as the gray (Gy), one gray being equal to the absorption of one joule of energy by one kilogram of food. The energy possessed by an electron is called an electron volt (eV). One eV is the amount of kinetic energy gained by an electron as it accelerates through an electric potential difference of one volt. It is usually more convenient to use a larger unit such as megaelectron volt (MeV), which is equal to one million electron volts.

Biological Effects of Irradiation

Irradiation has both direct and indirect effects on biological materials. The direct effects are due to the collision of radiation with atoms, resulting in an ejection of electrons from the atoms. The indirect effects are due to the formation of free radicals (unstable molecules carrying an extra electron) during the radiolysis (radiation-induced splitting) of water molecules. The radiolysis of water molecules produces hydroxyl radicals, highly reactive species that interact with the organic molecules present in foods. The products of these interactions cause many of the characteristics associated with the spoilage of food, such as off-flavours and off-odours.

Positive Effects

The bactericidal (bacteria-killing) effect of ionizing radiation is due to damage of the biomolecules of bacterial cells. The free radicals produced during irradiation may destroy or change the structure of cellular membranes. In addition, radiation causes irreversible changes to the nucleic acid molecules (i.e., DNA and RNA) of bacterial cells, inhibiting their ability to grow. Pathogenic bacteria that are unable to produce resistant endospores in foods such as poultry, meats, and seafood can be eliminated by radiation doses of 3 to 10 kilograys. If the dose of radiation is too low, then the damaged DNA can be repaired by specialized enzymes. If oxygen is present during irradiation, the bacteria are more readily damaged. Doses in the range of 0.2 to 0.36 kilograys are required to stop the reproduction of Trichinella spiralis (the parasitic worm that causes trichinosis) in pork, although much higher doses are necessary to eliminate it from the meat.

The dose of radiation used on food products is divided into three levels. Radappertization is a dose in the range of 20 to 30 kilograys, necessary to sterilize a food product. Radurization is a dose of 1 to 10 kilograys, that, like pasteurization, is useful for targeting specific pathogens. Radicidation involves doses of less than 1 kilogray for extending shelf life and inhibiting sprouting.

Negative Effects

In the absence of oxygen, radiolysis of lipids leads to cleavage of the interatomic bonds in the fat molecules, producing compounds such as carbon dioxide, alkanes, alkenes, and aldehydes. In addition, lipids are highly vulnerable to oxidation by free radicals, a process that yields peroxides, carbonyl compounds, alcohols, and lactones. The consequent rancidity, resulting from the irradiation of high-fat foods, is highly destructive to their sensory quality. To minimize such harmful effects, fatty foods must be vacuum-packaged and held at subfreezing temperatures during irradiation.

Proteins are not significantly degraded at the low doses of radiation employed in the food industry. For this reason irradiation does not inactivate enzymes involved in food spoilage, as most enzymes survive doses of up to 10 kilograys. On the other hand, the large carbohydrate molecules that provide structure to foods are depolymerized (broken down) by irradiation. This depolymerization reduces the gelling power of the long chains of structural carbohydrates. However, in most foods some protection against these deleterious effects is provided by other food constituents. Vitamins A, E, and B1 (thiamine) are also sensitive to irradiation. The nutritional losses of a food product are high if air is not excluded during irradiation.

Safety Concerns

Based on the beneficial effects of irradiation on certain foods, several countries have permitted its use for specific purposes, such as the inhibition of sprouting of potatoes, onions, and garlic; the extension of shelf life of strawberries, mangoes, pears, grapes, cherries, red currants, and cod and haddock fillets; and the insect disinfestation of pulses, peanuts, dried fruits, papayas, wheat, and ground-wheat products.

The processing room used for irradiation of foods is lined with lead or thick concrete walls to prevent radiation from escaping. The energy source, such as a radioactive element or a machine source of electrons, is located inside the room. (Radioactive elements such as ^{60}Co are contained in stainless steel tubes. Because an isotope cannot be switched on or off, when not in use it is lowered into a large reservoir of water.) Prior to the irradiation treatment, personnel vacate the room. The food to be irradiated is then conveyed by remote means into the room and exposed to the radiation source for a predetermined time. The time of exposure and the distance between the radiation source and the food material determine the irradiation treatment. After treatment, the irradiated food is conveyed out of the room, and the radioactive element is again lowered into the water reservoir.

Large-scale studies conducted around the world have concluded that irradiation does not cause harmful reactions in foods. In 1980 a joint committee of the Food and Agriculture Organization (FAO), the International Atomic Energy Agency (IAEA), and the World Health Organization (WHO) declared that an overall average dose of radiation of 10 kilograys was safe for food products. The maximum energy emitted by 60Co and 137Cs is too low to induce radioactivity in food. The energy output of electron-beam generators is carefully regulated, and the recommended energy outputs are too low to cause radioactivity in foods.

BRINING

In food processing, brining is treating food with brine or coarse salt which preserves and seasons the food while enhancing tenderness and flavor with additions such as herbs, spices, sugar, caramel and vinegar. Meat and fish are typically brined for less than twenty-four hours while vegetables, cheeses and fruit are brined in a much longer process known as pickling. Brining is similar to marination, except that a marinade usually includes a significant amount of acid, such as vinegar or citrus juice. Brining is also similar to curing, which usually involves significantly drying the food, and is done over a much longer time period.

Meat

Brining is typically a process in which meat is soaked in a salt water solution similar to marination before cooking. Meat is soaked anywhere from 30 minutes to several days. The amount of time needed to brine depends on the size of the meat: More time is needed for a large turkey compared to a broiler fryer chicken. Similarly, a large roast must be brined longer than a thin cut of meat.

Dry Brining

Brining can also be achieved by covering the meat in dry coarse salt and left to rest for several hours. The salt draws moisture from the interior of the meat to the surface, where it mixes with the salt and is then reabsorbed with the salt essentially brining the meat in its own juices. The salt rub is then rinsed off and discarded before cooking.

Kitchen salt applied to chicken showing extracted moisture after one hour.

Food scientists have two theories about the brining effect, but which one is correct is still under debate.

- The brine surrounding the cells has a higher concentration of salt than the fluid within the cells, but the cell fluid has a higher concentration of other solutes. This leads salt ions to diffuse into the cell, while the solutes in the cells cannot diffuse through the cell membranes into the brine. The increased salinity of the cell fluid causes the cell to absorb water from the brine via osmosis.

- The salt introduced into the cell denatures its proteins. The proteins coagulate, forming a matrix that traps water molecules and holds them during cooking. This prevents the meat from dehydrating.

Fish

Brined herring.

As opposed to dry salting, fish brining or wet-salting is performed by immersion of fish into brine, or just sprinkling it with salt without draining the moisture. To ensure long-term preservation, the solution has to contain at least 20% of salt, a process called "heavy salting" in fisheries; heavy-salted fish must be desalted in cold water or milk before consumption. If less salt is used, the fish is suited for immediate consumption, but additional refrigeration is necessary for longer preservation.

Wet-salting is used for preparation of:

- Salted herring, non-gutted, with hard or soft roe and heavily salted (20% NaCl brine, with final product containing around 12% salt),

- Soused herring which is gutted and lightly salted (2–3% NaCl), without roe,

- Anchovies, which can be immersed in brine or wet-salted. After several years, the fish liquifies and can be processed into paste or anchovy butter,

- Caviar and other types of roe.

Vegetables

Cucumbers in brine (dill pickles).

Pickled vegetables are immersed in brine, vinegar or vinaigrette for extended periods of time, where they undergo anaerobic fermentation which affects their texture and flavor. Pickling can preserve perishable foods for months. Antimicrobial herbs and spices, such as mustard seed, garlic, cinnamon or cloves, are often added. Unlike the canning process, pickling (which includes fermentation) does not require that the food be completely sterile before it is sealed. The acidity or salinity of the solution, the temperature of fermentation, and the exclusion of oxygen determine which microorganisms dominate, and determine the flavor of the end product.

Cheese

Brine is also commonly used to age brined cheeses, such as halloumi and feta; others are periodically washed in brine during their ripening. Not only does the brine carry flavors into the cheese (it might be seasoned with spices or wine), but the salty environment may nurture the growth of the *Brevibacterium linens* bacteria, which can impart a very pronounced odor (e.g. in Limburger) and interesting flavor. The same bacteria can also have some effect on cheeses that are simply ripened in humid conditions, like Camembert. Large populations of these "smear bacteria" show up as a sticky orange-red layer on some brine washed cheeses.

SMOKING

Smoking is the process of flavoring, browning, cooking, or preserving food by exposing it to smoke from burning or smoldering material, most often wood. Meat, fish, and *lapsang souchong* tea are often smoked.

In Europe, alder is the traditional smoking wood, but oak is more often used now, and beech to a lesser extent. In North America, hickory, mesquite, oak, pecan, alder, maple, and fruit-tree woods, such as apple, cherry, and plum, are commonly used for smoking. Other biomass besides wood can also be employed, sometimes with the addition of flavoring ingredients. Chinese tea-smoking uses a mixture of uncooked rice, sugar, and tea, heated at the base of a wok.

Meat hanging inside a smokehouse in Switzerland.

Some North American ham and bacon makers smoke their products over burning corn-cobs. Peat is burned to dry and smoke the barley malt used to make whisky and some beers. In New Zealand, sawdust from the native manuka (tea tree) is commonly used for hot smoking fish. In Iceland, dried sheep dung is used to cold-smoke fish, lamb, mutton and whale.

Hot-smoked chum salmon.

Historically, farms in the Western world included a small building termed the "smoke-house", where meats could be smoked and stored. This was generally well-separated from other buildings both because of the fire danger and because of the smoke emana-tions; the smoking of food could possibly introduce polycyclic aromatic hydrocarbons which may lead to an increased risk of some types of cancer. However, this association is still being debated.

A Montreal smoked meat sandwich.

Smoking can be done in four ways: cold smoking, warm smoking, hot smoking, and through the employment of "liquid smoke". However, these methods of imparting smoke only affect the food surface, and are unable to preserve food, thus, smoking is paired with other microbial hurdles, such as chilling and packaging, to extend food shelf-life.

Types

A home smoker and racks with hot smoked Pacific salmon.

Cold Smoking

Cold smoking differs from hot smoking in that the food remains raw, rather than cooked, throughout the smoking process. Smokehouse temperatures for cold smoking are typically done between 20 to 30 °C (68 to 86 °F). In this temperature range, foods take on a smoked flavor, but remain relatively moist. Cold smoking does not cook foods, and as such, meats should be fully cured before cold smoking. Cold smoking can be used as a flavor enhancer for items such as cheese or nuts, along with meats such as chicken breasts, beef, pork chops, salmon, scallops, and steak. The item is often hung in a dry environment first to develop a pellicle, then it can be cold smoked up to several days to ensure it absorbs the smokey flavour. Some cold smoked foods are baked, grilled, steamed, roasted, or sautéed before eating.

Cold smoking meats is not something that should be attempted at home, according to the US National Center for Home Food Preservation: "Most food scientists cannot recommend cold-smoking methods because of the inherent risks." Cold smoking meats should only be attempted by personnel certified in HACCP, or Hazard Analysis and Critical Control Points, to ensure that it is safely prepared.

Warm Smoking

Warm smoking exposes foods to temperatures of 25–40 °C (77–104 °F).

Hot Smoking

Hot smoking exposes the foods to smoke and heat in a controlled environment such as a smoker oven or smokehouse. Hot smoking requires the use of a smoker which generates heat either from a charcoal base, heated element within the smoker or from a stove-top or oven; food is hot smoked by cooking and flavoured with wood smoke simultaneously. Like cold smoking, the item may be hung first to develop a pellicle; it is then smoked from 1 hour to as long as 24 hours. Although foods that have been hot smoked are often reheated or further cooked, they are typically safe to eat without further cooking. Hams and ham hocks are fully cooked once they are properly smoked, and they can be eaten as is without any further preparation. Hot smoking usually occurs within the range of 52 to 80 °C (126 to 176 °F). When food is smoked within this temperature range, foods are fully cooked, moist, and flavorful. If the smoker is allowed to get hotter than 85 °C (185 °F), the foods will shrink excessively, buckle, or even split. Smoking at high temperatures also reduces yield, as both moisture and fat are cooked away.

Liquid Smoke

Liquid smoke, a product derived from smoke compounds in water, is applied to foods through spraying or dipping.

Smoke Roasting

Smoke-roasting refers to any process that has the attributes of both roasting and smoking. This smoking method is sometimes referred to as barbecuing or pit-roasting. It may be done in a smoke-roaster, a closed wood-fired oven, or a barbecue pit, any smoker that can reach above 121 °C (250 °F), or in a conventional oven (one that a person does not mind having smoky all the time) by placing a pan filled with hardwood chips on the floor of the oven so that the chips can smolder and produce a smoke-bath. In North America, this smoking method is commonly referred to as "barbecuing," "pit baking," or "pit roasting".

Wood Smoke

Hickory-smoked country-style ribs.

Hardwoods are made up mostly of three materials: cellulose, hemicellulose, and lignin. Cellulose and hemicellulose are the basic structural material of the wood cells; lignin acts as a kind of cell-bonding glue. Some softwoods, especially pines and firs, hold significant quantities of resin, which produces a harsh-tasting soot when burned; these woods are not often used for smoking.

Cellulose and hemicellulose are aggregate sugar molecules; when burnt, they effectively caramelize, producing carbonyls, which provide most of the color components and sweet, flowery, and fruity aromas. Lignin, a highly complex arrangement of interlocked phenolic molecules, also produces a number of distinctive aromatic elements when burnt, including smoky, spicy, and pungent compounds such as guaiacol, phenol, and syringol, and sweeter scents such as the vanilla-scented vanillin and clove-like isoeugenol. Guaiacol is the phenolic compound most responsible for the "smoky" taste, while syringol is the primary contributor to smoky aroma. Wood also contains small quantities of proteins, which contribute roasted flavors. Many of the odor compounds in wood smoke, especially the phenolic compounds, are unstable, dissipating after a few weeks or months.

A number of wood smoke compounds act as preservatives. Phenol and other phenolic compounds in wood smoke are both antioxidants, which slow rancidification of animal fats, and antimicrobials, which slow bacterial growth. Other antimicrobials in wood smoke include formaldehyde, acetic acid, and other organic acids, which give wood smoke a low pH—about 2.5. Some of these compounds are toxic to people as well, and may have health effects in the quantities found in cooking applications.

Since different species of trees have different ratios of components, various types of wood do impart a different flavor to food. Another important factor is the temperature at which the wood burns. High-temperature fires see the flavor molecules broken down further into unpleasant or flavorless compounds. The optimal conditions for smoke flavor are low, smoldering temperatures between 300 and 400 °C (570 and 750 °F). This is the temperature of the burning wood itself, not of the smoking environment, which uses much lower temperatures. Woods that are high in lignin content tend to burn hot; to keep them smoldering requires restricted oxygen supplies or a high moisture content. When smoking using wood chips or chunks, the combustion temperature is often raised by soaking the pieces in water before placing them on a fire.

Types of Smokers

Offset

An example of a common offset smoker.

The main characteristics of the offset smoker are that the cooking chamber is usually cylindrical in shape, with a shorter, smaller diameter cylinder attached to the bottom of one end for a firebox. To cook the meat, a small fire is lit in the firebox, where airflow is tightly controlled. The heat and smoke from the fire are drawn through a connecting pipe or opening into the cooking chamber.

The heat and smoke cook and flavor the meat before escaping through an exhaust vent at the opposite end of the cooking chamber. Most manufacturers' models are based on this simple but effective design, and this is what most people picture when they think of a "BBQ smoker". Even large capacity commercial units use this same basic design of a separate, smaller fire box and a larger cooking chamber.

Upright Drum

The upright drum smoker (also referred to as an ugly drum smoker or UDS) is exactly what its name suggests; an upright steel drum that has been modified for the purpose of pseudo-indirect hot smoking. There are many ways to accomplish this, but the basics include the use of a complete steel drum, a basket to hold charcoal near the bottom, and cooking rack (or racks) near the top; all covered by a vented lid of some sort. They have been built using many different sizes of steel drums, such as 30 US gallons (110 l; 25 imp gal), 55 US gallons (210 l; 46 imp gal), and 85 US gallons (320 l; 71 imp gal) for example, but the most popular size is the common 55-gallon drum.

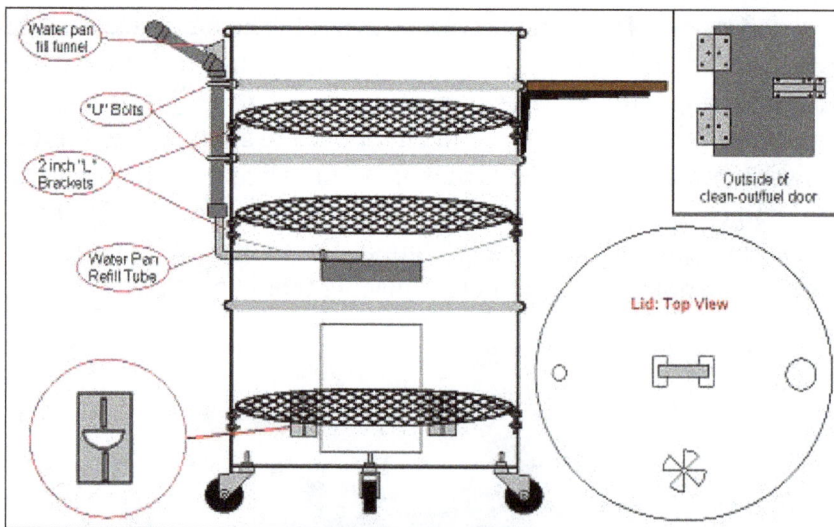

A diagram of a typical upright drum smoker.

This design is similar to smoking with indirect heat due to the distance from the coals and the racks, which is typically 24 inches (61 cm). The temperatures used for smoking are controlled by limiting the amount of air intake at the bottom of the drum, and allowing a similar amount of exhaust out of vents in the lid. UDSs are very efficient with fuel consumption and flexible in their abilities to produce proper smoking conditions, with or without the use of a water pan or drip pan. Most UDS builders/users would say a water pan defeats the true pit BBQ nature of the UDS, as the drippings from the smoked meat should land on the coals, burning up, and imparting a unique flavor one cannot get with a water pan.

Vertical Water Smoker

A typical vertical water smoker.

A vertical water smoker (also referred to as a bullet smoker because of its shape) is a variation of the upright drum smoker. It uses charcoal or wood to generate smoke and heat, and contains a water bowl between the fire and the cooking grates. The water bowl serves to maintain optimal smoking temperatures and also adds humidity to the smoke chamber. It also creates an effect in which the water vapor and smoke condense together, which adds flavor to smoked foods. In addition, the bowl catches any drippings from the meat that may cause a flare-up. Vertical water smokers are extremely temperature stable and require very little adjustment once the desired temperature has been reached. Because of their relatively low cost and stable temperature, they are sometimes used in barbecue competitions where propane and electric smokers are not allowed.

Propane Smoker

A propane smoker is designed to allow the smoking of meat in a somewhat more temperature controlled environment. The primary differences are the sources of heat and of the smoke. In a propane smoker, the heat is generated by a gas burner directly under a steel or iron box containing the wood or charcoal that provides the smoke. The steel box has few vent holes, on the top of the box only. By starving the heated wood of oxygen, it smokes instead of burning. Any combination of woods and charcoal may used. This method uses much less wood but does require propane fuel.

A diagram of a propane smoker, loaded with
country style ribs and pork loin in foil.

Smoke Box

This more traditional method uses a two-box system: a fire box and a food box. The fire box is typically adjacent or under the cooking box, and can be controlled to a finer degree. The heat and smoke from the fire box exhausts into the food box, where it is used to cook and smoke the meat. These may be as simple as an electric heating element with a pan of wood chips placed on it, although more advanced models have finer temperature controls.

Electric Smokers

The most convenient of the various types of smokers are the insulated electric smokers. These devices house a heating element that can maintain temperatures ranging from that required for a cold smoke all the way up to 135 °C (275 °F) with little to no intervention from the user. Although wood chunks, pellets, and even in some cases automatically-fed wood pucks are used to generate smoke, the amount of flavor obtained is less than traditional wood or charcoal smokers.

Trench

In this method the firebox is a narrow trench cut down a slope pointing into the prevailing wind. The middle part of the trench is covered over to make it into a tunnel. At the upper end of the trench is a vertical framework covered to form a chimney within which is placed the rack of foodstuff. At the lower upwind end of the trench is lit a small smokey fire, and sustained day and night until the foodstuff is cured.

Commercial Smokehouse

Commercial smokehouses, mostly made from stainless steel, have independent systems for smoke generation and cooking. Smoke generators use friction, an electric coil or a small flame to ignite sawdust on demand. Heat from steam coils or gas flames is balanced with live steam or water sprays to control the temperature and humidity. Elaborate air handling systems reduce hot or cold spots, to reduce variation in the finished product. Racks on wheels or rails are used to hold the product and facilitate movement.

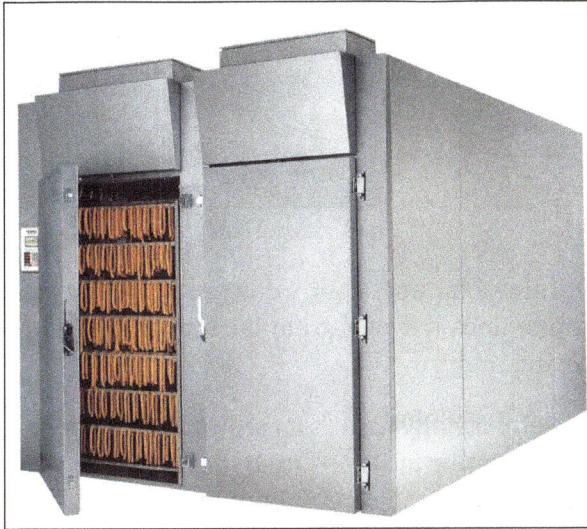

Commercial Smokehouse.

Pellet Smokers

A pellet smoker is a temperature controlled smoker that burns wood pellets made of dried out sawdust, about an inch long and 1/4 inch wide. The wood pellets are stored in a gravity-fed hopper that feeds into a motor controlled by the temperature regulator. This motor pushes the pellets into an auger that sits underneath the heat box. An ignition rod within the auger ignites the pellets where a combustion fan keeps them smouldering. The motor and the combustion fan regulate the temperature of the smoker by feeding it more pellets and increasing airflow in the auger. Above the auger is a heat shield to disperse the direct heat before it reaches the heat box to allow the wood smoke to keep the heat box at an even temperature throughout. The heat sensor inside the heat box relays the current temperature inside the box back to the temperature regulator which then controls the fan speed and pellet hopper motor which will either increase or decrease the amount of pellets in the auger or the amount of air available to the fire to maintain the desired temperature for the cook.

The popularity of this type of smokers is on the rise after many BBQ pit-masters started using them for competition barbeque.

Preservation

Smoked omul fish.

Smoke is both an antimicrobial and antioxidant, however it is insufficient alone for preserving food as smoke does not penetrate far into meat or fish; it is thus typically combined with salt-curing or drying.

Smoking is especially useful for oily fish, as its antioxidant properties inhibit surface fat rancidification and delay interior fat exposure to degrading oxygen. Some heavily-salted, long-smoked fish can keep without refrigeration for weeks or months.

Artificial smoke flavoring (such as liquid smoke) can be purchased to mimic smoking's flavor, but not its preservative qualities.

Competitive Smoking

Competition BBQ Smoking is becoming increasingly popular among smoking enthusiasts, especially in the Southern American States, where BBQ enthusiasts come together to over a weekend to cook various cuts of meat such as a whole hog or beef brisket to become the best at BBQ.

List of Smoked Foods and Beverages

Beverages

- *Lapsang souchong* tea leaves are smoked and dried over pine or cedar fires.
- Malt beverages:
  - The malt used to make whisky.
  - Rauchbier (smoked beer).

Smoked Gruyère cheese.

Fish being smoked.

Fruit and Vegetables

- Capsicums: Chipotles (smoked, ripe jalapeños), paprika.
- Prunes (dried plums) can be smoked while drying.
- *Wumei* are smoked plum fruits.
- *Iburi-gakko* are a smoked daikon pickle from Akita Prefecture, Japan.

Meat, Fish and Cheese

- Beef:
  - Pastrami (pickled, spiced and smoked beef brisket).
- Pork:
  - Bacon.
  - Ham.
  - Bakkwa.
- Turkey.
- Sausage:
  - Salami.
- Jerky.
- Fish:
  - Eel popular in eastern/northern Europe.
  - Traditional Grimsby smoked fish (cod and haddock).
  - Haddock and Arbroath Smokies (haddock).

- ○ Buckling, kippers and bloater (herring).

- ○ Salmon.

- ○ Mackerel.

- Egg (eggs and fish eggs).

- Cheese:

  - ○ Gouda.

  - ○ Gruyère.

Other Proteins

- Nuts.

- Tofu.

Spices

- Paprika.

- Salt.

FOOD PRESERVATION BY EXTRUSION

Food extrusion is a form of extrusion used in food processing. It is a process by which a set of mixed ingredients are forced through an opening in a perforated plate or die with a design specific to the food, and are then cut into a specific size by blades. The machine which forces the mix through the die is an extruder, and the mix is known as the extrudate. The extruder consists of a large, rotating screw tightly fitting within a stationary barrel, at the end of which is the die.

Extrusion enables mass production of food via a continuous, efficient system that ensures uniformity of the final product. Food products manufactured using extrusion usually have a high starch content. These include some pasta, breads (croutons, bread sticks, and flat breads), many breakfast cereals and ready-to-eat snacks, confectionery, pre-made cookie dough, some baby foods, full-fat soy, textured vegetable protein, some beverages, and dry and semi-moist pet foods.

Process

In the extrusion process, raw materials are first ground to the correct particle size, usually the consistency of coarse flour. The dry mix is passed through a pre-conditioner, in which other ingredients are added depending on the target product; these may be liquid sugar, fats, dyes, meats or water. Steam is injected to start the cooking process, and the preconditioned mix (extrudate) is then passed through an extruder. The extruder consists of a large, rotating screw tightly fitting within a stationary barrel, at the end of which is the die. The extruder's rotating screw forces the extrudate toward the die, through which it then passes. The amount of time the extrudate is in the extruder is the residence time.

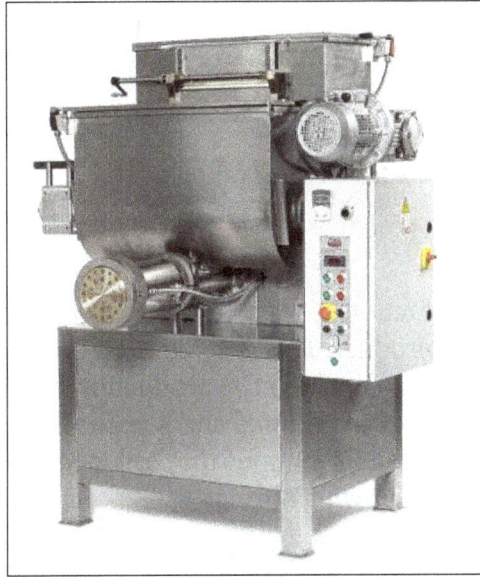

A non-vacuum short goods pasta extruder.

The extruded product usually puffs and changes texture as it is extruded because of the reduction of forces and release of moisture and heat. The extent to which it does so is known as the expansion ratio. The extrudate is cut to the desired length by blades at the output of the extruder, which rotate about the die openings at a specific speed. The product is then cooled and dried, becoming rigid while maintaining porosity.

The cooking process takes place within the extruder where the product produces its own friction and heat due to the pressure generated (10–20 bar). The process can induce both protein denaturation and starch gelatinization under some conditions.

Many food extrusion processes involve a high temperature over a short time. Important factors of the extrusion process are the composition of the extrudate, screw length and rotating speed, barrel temperature and moisture, die shape, and rotating speed of the blades. These are controlled based on the desired product to ensure uniformity of the output.

Moisture is the most important of these factors, and affects the mix viscosity, acting to plasticize the extrudate. Increasing moisture will decrease viscosity, torque, and product temperature, and increase bulk density. This will also reduce the pressure at the die. Most extrusion processes for food processing maintain a moisture level below 40%, that is low to intermediate moisture. High-moisture extrusion is known as wet extrusion, but it was not used much before the introduction of twin screw extruders (TSE), which have a more efficient conveying capability. The most important rheological factor in the wet extrusion of high-starch extrudate is temperature.

The amount of salt in the extrudate may determine the colour and texture of some extruded products. The expansion ratio and airiness of the product depend on the salt concentration in the extrudate, possibly as a result of a chemical reaction between the salt and the starches in the extrudate. Colour changes as a result of salt concentration may be caused by "the ability of salt to change the water activity of the extrudate and thus change the rate of browning reactions". Salt is also used to distribute minor ingredients, such as food colours and flavours, after extrusion; these are more evenly distributed over the product's surface after being mixed with salt.

Effects

Extrusion enables mass production of food via a continuous, efficient system that ensures uniformity of the final product. This is achieved by controlling various aspects of the extrusion process. It has also enabled the production of new processed food products and "revolutionized many conventional snack manufacturing processes". The extrusion process results in "chemical reactions that occur within the extruder barrel and at the die". Extrusion has the following effects:

- Destruction of certain naturally occurring toxins.

- Reduction of microorganisms in the final product.

- Slight increase of iron-bioavailability.

- Creation of insulin-desensitizing starches (a potential risk-factor for developing diabetes).

- Loss of lysine, an essential amino acid necessary for developmental growth and nitrogen management.

- Simplification of complex starches, increasing rates of tooth decay.

- Increase of glycemic index of the processed food, as the "extrusion process significantly increased the availability of carbohydrates for digestion".

- Destruction of Vitamin A (beta-carotene).

- Denaturation of proteins.

The material of which an extrusion die is made can affect the final product. Compared to stainless steel dies, a pasta machine with bronze dies produces a rougher surface. This is considered to give an improved taste, as it better retains pasta sauces. "Bronze die" pasta is labelled as such on retail packages, to indicate a premium product.

The effects of "extrusion cooking on nutritional quality are ambiguous", as extrusion may change carbohydrates, dietary fibre, the protein and amino acid profile, vitamins, and mineral content of the extrudate in a manner that is beneficial or harmful.

High-temperature extrusion for a short duration "minimizes losses in vitamins and amino acids". Extrusion enables mass production of some food, and will "denature antinutritional factors", such as destroying toxins or killing microorganisms. It may also improve "protein quality and digestibility", and affects the product's shape, texture, colour, and flavour.

It may also cause the fragmentation of proteins, starches, and non-starch polysaccharides to create "reactive molecules that may form new linkages not found in nature". This includes Maillard reactions which reduce the nutritional value of the proteins. Vitamins with heat lability may be destroyed. As of 1998, little is known about the stability or bioavailability of phytochemicals involved in extrusion. Nutritional quality has been found to improve with moderate conditions (short duration, high moisture, low temperature), whereas a negative effect on nutritional quality of the extrudate occurs with a high temperature (at least 200 °C), low moisture (less than 15%), or improper components in the mix.

Products

Extrusion has enabled the production of new processed food products and "revolutionized many conventional snack manufacturing processes".

Cheese curls made using an extruder.

The various types of food products manufactured by extrusion typically have a high starch content. *Directly expanded* types include breakfast cereals and corn curls, and are made in high temperature, low moisture conditions under high shear. *Unexpanded* products include pasta, which is produced at intermediate moisture (about 40%) and low temperature. *Texturized* products include meat analogues, which are made using plant proteins ("textured vegetable protein") and a long die to "impart a fibrous, meat-like structure to the extrudate", and fish paste. Confectionery made via extrusion includes chewing gum, liquorice, and toffee.

Some processed cheeses and cheese analogues are also made by extrusion. Processed cheeses extruded with low moisture and temperature "might be better suited for manufacturing using extrusion technology" than those at high moisture or temperature. Lower moisture cheeses are firmer and chewier, and cheddar cheese with low moisture and an extrusion temperature of 80 °C was preferred by subjects in a study to other extruded cheddar cheese produced under different conditions. An extrudate mean residence time of about 100 seconds can produce "processed cheeses or cheese analogues of varying texture (spreadable to sliceable)".

Other food products often produced by extrusion include some breads (croutons, bread sticks, and flat breads), various ready-to-eat snacks, pre-made cookie dough, some baby foods, some beverages, and dry and semi-moist pet foods. Specific examples include cheese curls, macaroni, Fig Newtons, jelly beans, sevai, and some french fries. Extrusion is also used to modify starch and to pellet animal feed.

FOOD PRESERVATION BY CANNING

Canning is a method of preserving food by first sealing it in air-tight jars, cans or pouches, and then heating it to a temperature that destroys contaminating microorganisms that can either be of health or spoilage concern because of the danger posed by several spore-forming thermo-resistant microorganisms, such as Clostridium botulinum (the causative agent of botulism). Spores of C. botulinum (in a concentration of 104/ml) can resist boiling at 100 °C (212 °F) for more than 300 minutes; however, as temperature increases the times decrease exponentially, so at 121 °C (250 °F) for the same concentration just 2.8 minutes are required.

Cans of preserved food.

From a public safety point of view, foods with low acidity (i.e., pH > 4.3) need sterilization by canning under conditions of both high temperature (116-130 °C) and pressure. Foods that must be pressure canned include most vegetables, meats, seafood, poultry, and dairy products. The only foods that may be safely canned in a boiling water bath (without high pressure) are highly acidic foods with a pH below 4.6, such as fruits, pickled vegetables, or other foods to which acid has been added.

Double Seams

Modern double seams provide an airtight seal to the tin can. This airtight nature is crucial to keeping bacteria out of the can and keeping its contents sealed inside. Thus, double seamed cans are also known as Sanitary Cans. Developed in 1900 in Europe, this sort of can was made of the traditional cylindrical body made with tin plate; however, the two ends (lids) were attached using what is now called a double seam. A can thus sealed is impervious to the outside world by creating two tight continuous folds between the can's cylindrical body and the lid at each end. This eliminated the need for solder and allowed improvements in the speed of manufacturing, thereby lowering the cost.

Double seams make extensive use of rollers in shaping the can, lid and the final double seam. To make a sanitary can and lid suitable for double seaming, manufacture begins with a sheet of coated tin plate. To create the can body rectangles are cut and curled around a die and welded together creating a cylinder with a side seam.

Rollers are then used to flare out one or both ends of the cylinder to create a quarter circle flange around the circumference. Great care and precision are required to ensure that the welded sides are perfectly aligned, as any misalignment will mean that the shape of the flange is inconsistent, compromising its integrity.

A circle is then cut from the sheet using a die cutter. The circle is shaped in a stamping press to create a downward countersink to fit snugly in to the can body. The result can be compared to an upside down and very flat top hat. The outer edge is then curled down and around approximately 130 degrees using rollers creating the end curl.

The final result is a steel tube with a flanged edge. And a countersunk steel disc with a curled edge. A rubber compound is put inside the curl.

Seaming

The body and end are brought together in a seamer and held in place by the base plate and chuck, respectively. The base plate provides a sure footing for the can body during the seaming operation and the chuck fits snugly in to the end (lid). The result is the countersink of the end sits inside the top of the can body just below the flange. The end curl protrudes slightly beyond the flange.

First Operation

Once brought together in the seamer, the seaming head presses a special first operation roller against the end curl. The end curl is pressed against the flange curling it in toward the body and under the flange. The flange is also bent downward and the end and body are now loosely joined together. The 1st operation roller is then retracted. At this point during manufacture five thicknesses of steel exist in the seam. From the outside in they are; a) End, b) Flange, c) End Curl, d) Body, e) Countersink. This is the first seam. All the parts of the seam are now aligned and ready for the final stage.

Second Operation

The seaming head then engages the second operation roller against the partly formed seam. The second operation presses all five steel components together tightly to form the final seal. The five layers in the final seam are then called; a) End, b) Body Hook, c) Cover Hook, d) Body, e) Countersink. All sanitary cans require a filling medium within the seam as metal to metal contact, otherwise such an arrangement would not maintain its hermetic seal for very long. In most cases a rubberized sealing compound is placed inside the end curl radius, forming the actual critical contact point between the end and the body.

Probably the most important innovation since the introduction of double seams is the welded side seam. Prior to the welded side seam the can body was folded and soldered together, leaving a relatively thick side seam. The thick side seam meant that at the side seam end juncture the end curl had more metal to curl around before closing in behind the Body Hook or flange, leaving a greater opportunity for error.

References

- Food-preservation: britannica.com, Retrieved 5 February, 2019

- Rahman, m. Shafiur, ed. (2007). Handbook of food preservation (2nd ed.). Boca raton: crc press. Isbn 9781420017373

- Preservatives: foodadditivesworld.com, Retrieved 26 July, 2019

- "Pickle bill fact sheet". 13 march 2008. Archived from the original on 13 march 2008. Retrieved 15 february2018

- Facts-6790531-meaning-thermal-processing: sciencing.com, Retrieved 21 May, 2019

- Klein, donald w.; lansing m.; harley, john (2006). Microbiology (6th ed.). New york: mcgraw-hill. Isbn 978-0-07-255678-0

- Thermal-processing-of-food: tiselab.com, Retrieved 8 January, 2019

- Benwick, bonnie s. (november 14, 2007). "wet brining vs. Dry: give that bird a bath". The washington post. Retrieved 2018-04-07

- Food-irradiation, food-preservation: britannica.com, Retrieved 13 May, 2019

High Pressure Processing of Food

High pressure food processing is a method of preservation and sterilization of food under high pressure. Such processing has numerous effects on food characteristics, food microorganisms and food components. The topics elaborated in this chapter will help in gaining a better perspective about high pressure food processing as well as its effects.

PASCALIZATION

Pascalization, bridgmanization, high pressure processing (HPP) or high hydrostatic pressure (HHP) processing is a method of preserving and sterilizing food, in which a product is processed under very high pressure, leading to the inactivation of certain microorganisms and enzymes in the food. HPP has a limited effect on covalent bonds within the food product, thus maintaining both the sensory and nutritional aspects of the product. The technique was named after Blaise Pascal, a French scientist of the 17th century whose work included detailing the effects of pressure on fluids. During pascalization, more than 50,000 pounds per square inch (340 MPa, 3.4 kbar) may be applied for around fifteen minutes, leading to the inactivation of yeast, mold, and bacteria. Pascalization is also known as bridgmanization, named for physicist Percy Williams Bridgman.

Uses

Spoilage microorganisms and some enzymes can be deactivated by HPP, which can extend the shelf life while preserving the sensory and nutritional characteristics of the product. Pathogenic microorganisms such as *Listeria, E. coli, Salmonella,* and *Vibrio* are also sensitive to pressures of 400-1000 MPa used during HPP. Thus, HPP can pasteurize food products with decreased processing time, reduced energy usage, and less waste. The treatment occurs at low temperatures and does not include the use of food additives. From 1990, some juices, jellies, and jams have been preserved using pascalization in Japan. The technique is now used there to preserve fish and meats, salad dressing, rice cakes, and yogurts. HPP is now being used to preserve fruit and vegetable smoothies and other products such as meat for sale in the UK. An early use of pascalization in the United States was to treat guacamole. It did not change the guacamole's taste, texture, or color, but the shelf life of the product increased to thirty days, from three days without the treatment. However, some treated foods still require cold storage because pascalization does not stop all enzyme activity caused by proteins, some of which affects shelf life. In recent years, HPP has also been used in the processing of raw pet food. Most commercial frozen and freeze-dried raw diets now go through post-packaging HPP treatment to destroy potential bacteria and viruses contaminants, with salmonella being one of the biggest concerns.

Process

In pascalization, food products are sealed and placed into a steel compartment containing a liquid, often water, and pumps are used to create pressure. The pumps may apply pressure constantly or intermittently. The application of high hydrostatic pressures (HHP) on a food product will kill many microorganisms, but the spores are not destroyed. Pascalization works especially well on acidic foods, such as yogurts and fruits, because pressure-tolerant spores are not able to live in environments with low pH levels. The treatment works equally well for both solid and liquid products.

Bacterial spores survive pressure treatment at ambient or chilled conditions. Researchers reported that pressure in combination with heat is effective in the inactivation of bacterial spores. The process is called pressure-assisted thermal sterilization. In 2009 and 2015, Food and Drug Administration (FDA) issued letters of no objection for two industrial petition for pressure-assisted thermal processing. At this time, there are no commercial low-acid products treated by PATP are available in the market.

During pascalization, the food's hydrogen bonds are selectively disrupted. Because pascalization is not heat-based, covalent bonds are not affected, causing no change in the food's taste. This means that HPP does not destroy vitamins, maintaining the nutritional value of the food. High hydrostatic pressure can affect muscle tissues by increasing the rate of lipid oxidation, which in turn leads to poor flavor and decreased health benefits. Additionally, there are some compounds present in foods that are subject to change during the treatment process. For example, carbohydrates are gelatinized by an increase in pressure instead of increasing the temperature during the treatment process.

Because hydrostatic pressure is able to act quickly and evenly on food, neither the size of a product's container nor its thickness play a role in the effectiveness of pascalization. There are several side effects of the process, including a slight increase in a product's sweetness, but pascalization does not greatly affect the nutritional value, taste, texture, and appearance. As a result, high pressure treatment of foods is regarded as a "natural" preservation method, as it does not use chemical preservatives.

EFFECTS OF HIGH PRESSURE ON FOOD CHARACTERISTICS

Colour, flavour and texture are important quality characteristics of fruits and vegetables and major factors affecting sensory perception and consumer acceptance of foods. High hydrostatic pressure processing (HPP) is an emerging nonthermal technology that can ensure the same level of food safety as heat pasteurization and produces freshertasting, minimally processed foods. This technology reportedly increases shelf life, while minimizing loss of quality. Additionally, it maintains the nutritional value and quality of food and therefore does not result in any undesirable changes associated with thermal processing. Various processing methods are used not only to increase the edibility and palatability of fruits and vegetables but also to prolong their shelf life. Possible impacts of HP treatments at elevated temperatures on sensory properties are highlighted since the temperature regime used for research on high pressure (HP) has been extended to elevated temperatures in order to achieve spore inactivation.

With non-thermal processing technologies, more fresh-like products can be obtained. HHP is considered to be an alternative to thermal pasteurization for fruit juices and other products when this process is used alone or in combination with traditional techniques. The major benefit of pressure is its immediate and uniform effect throughout different media, avoiding difficulties such as nonstationary conditions typical for convection and conduction processes. Colour, flavour and texture are important quality characteristics of fruits and vegetables and major factors affecting sensory perception and consumer acceptance of foods. HP processing could preserve nutritional value and the delicate sensory properties of fruits and vegetables due to its limited effect on the covalent bonds of low molecularmass compounds such as colour and flavour compounds. However, food is a complex system and the compounds responsible for sensory properties coexist with enzymes, metal ions, etc. During HP processing, different pressure and temperature combinations can be used to achieve desired effects on texture, colour and flavour of foods. Covalent bonds in food are usually less affected during HPP, because the compression energy involved is low. Pressure alters interatomic distances, acting mainly on weak interactions in which bond energy is distance-dependent, such as van der Vaals forces, electrostatic forces, hydrogen bonding and hydrophobic interactions of proteins. However, these effects on hydrogen bonding and hydrophobic interaction are complicated depending on the structural properties of the considered compounds. The quality of HP processed fruits and vegetables can, however, change during storage due to coexisting chemical reactions, such as oxidation, and biochemical reactions when endogenous enzymes or microorganisms are incompletely inactivated.

Effect of HP Processing on Flavour

Any changes in the compounds responsible for the sourness, sweetness, bitterness or odour of fruits and vegetables may result in changes in their flavour. Flavour is the sensory impression of a food that is determined mainly by the chemical senses of taste and smell. Chiao et al., evaluated the effects of high-pressure treatment on microbial growth and production of off-flavor compounds in raw octopus during 16 days of refrigerated storage. Chopped raw octopus samples were treated at 150, 300, 450, and 600 MPa for 6 min using high-pressure laboratory food processing equipment. The number of psychrotrophic bacteria on day 16 was reduced by 0.1, 0.5, 1.3, and 2.8 CFU/g after pressure treatment at 150, 300, 450, and 600 MPa, respectively, as compared with control group. The amounts of trimethylamine (TMA) and dimethylamine (DMA) produced in the chopped raw octopus treated at 600 MPa was significantly reduced as compared to the levels in the control. The production of biogenic amines (BAs) increased up to 1.82 mg/g in the control after 12 days of refrigerated storage, while the BA levels in the 450 MPa- and 600 MPa-treated octopus were 1.40 and 1.35 mg/g, respectively. High-pressure treatment is a promising alternative technology for extending the shelf life of raw octopus.

The human tongue can distinguish only among five distinct qualities of taste, of which sourness, sweetness and bitterness are the most important ones regarding the flavour of fruits and vegetables. The human nose, on the other hand, can distinguish among a vast number of volatile compounds, even in minute quantities.

It is generally assumed that the fresh flavour of fruits and vegetables is not altered by high-pressure processing, since the structure of small molecular flavour compounds is not directly affected by high pressure. As HP processing can enhance and retard enzymatic and chemical reactions, it could indirectly alter the content of some flavour compounds and disturb the whole balance of flavour composition in fruits and vegetables. As a consequence, HP processing could result in undesired changes in flavour. Hexanal is a volatile compound associated with the smell of foliage and grass. Gas chromatographic studies showed changes in the hexanal content of fruits and vegetables as a result of HP processing. At concentrations lower than about 1.2 mg kg^{-1}, n-hexanal contributes to the typical fresh flavour of tomatoes. Higher concentrations impart a rancid flavour. The increased concentration of nhexanal was considered to be a result of HP-induced oxidation of free fatty acids, such as linoleic and linolenic acid.

Kyung et al., investigated the characterization of flavor, physicochemical properties and biological activities of garlic extracts prepared by high hydrostatic pressure (HHP) treatment (500 MPa) was conducted at various HHP reaction times and pH conditions. The evaluation of flavor revealed that HHP treated garlic samples in acidic condition (pH 1.8-3) were most effective to reduce the pungent flavor of garlic among all conditions. The antioxidative, antimicrobial and antitumor activities of HHP treated garlic samples were decreased compared with control. A rapid decrease in antimicrobial and antioxidative activities was observed over 3 min HHP reaction time. No antitumor activities were observed after 3 min HHP reaction time. Up to 56 s HHP reaction time, the alliinase activity was not changed significantly but it was dramatically decreased at a longer HHP reaction time compared with control (P < 0.05), showing higher stability in acidic condition than alkaline condition.

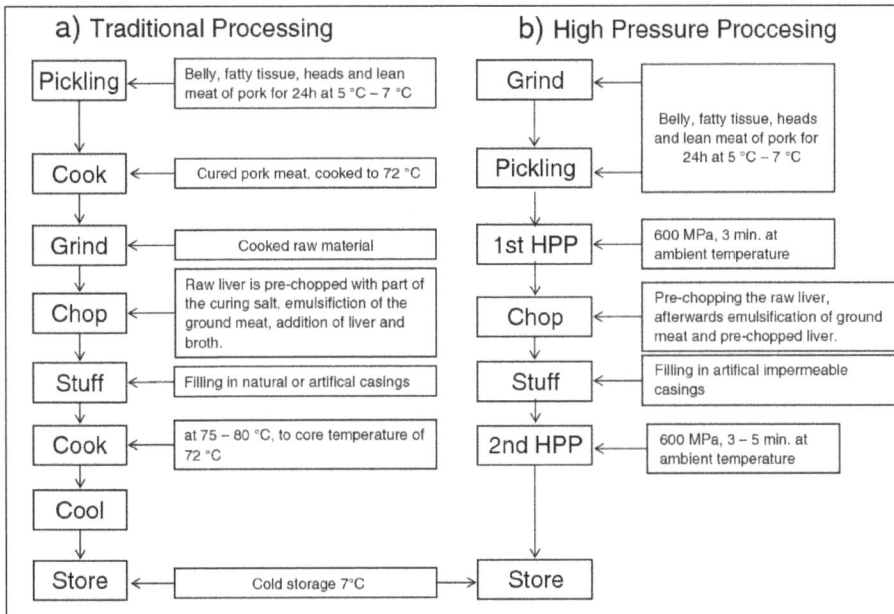

a) Traditional Processing		b) High Pressure Proccesing	
Pickling	Belly, fatty tissue, heads and lean meat of pork for 24h at 5 °C – 7 °C	Grind	
Cook	Cured pork meat, cooked to 72 °C	Pickling	Belly, fatty tissue, heads and lean meat of pork for 24h at 5 °C – 7 °C
Grind	Cooked raw material	1st HPP	600 MPa, 3 min. at ambient temperature
Chop	Raw liver is pre-chopped with part of the curing salt. emulsifiction of the ground meat, addition of liver and broth.	Chop	Pre-chopping the raw liver, afterwards emulsification of ground meat and pre-chopped liver.
Stuff	Filling in natural or artifical casings	Stuff	Filling in artifical impermeable casings
Cook	at 75 – 80 °C, to core temperature of 72 °C	2nd HPP	600 MPa, 3 – 5 min. at ambient temperature
Cool			
Store	Cold storage 7°C	Store	

Comparison of the traditional process: (a) and the novel high pressure processing, (b) for the production of German liver sausages.

Lipoxygenase and hydroperoxide lyase, which are naturally present in tomato, are partly responsible for the development of the rancid taste as they catalyse the oxidation of poly unsaturated fatty acids. At pressures lower than 500 MPa (20 °C), tomato hydroperoxide lyase is more labile than tomato lipoxygenase, while their stabilities are opposite at pressure 500 MPa. In diced tomatoes, lipoxygenase activity was reduced by almost 50% as a result of HP treatment at 400 MPa (25 or 45 °C/1-5 min) and it was very low after treatment at 800 MPa (25 or 45 °C/1 min). Regarding strawberry based food products, HP processing at 800 MPa (20 °C/20 min) modified the flavour profile of strawberry purée. Some new compounds were formed, e.g. glactone, which correlates with the flavour of peach. The concentration of many volatile compounds contributing to fresh strawberry flavour, such as nerolidol, furaneol, linalool and some ester compounds was significantly lower in the strawberry purée processed at 800 MPa (20 °C/20 min) than in the unprocessed purée.

After cold storage (1 day, 4 °C), the concentrations of acids (butanoic acid, 2-methyl-butanoic acid and hexanoic acid) and the ketone compound 2,4,6-heptanetrione of HP-treated (200, 400, 600 or 800 MPa/18-22 °C/15 min) strawberries were lower than in the untreated strawberries. The concentration of the alcohol 1,6,10-dodecatrien3-ol increased in strawberries treated at 800 MPa. Ester compounds belong to the most important flavor compounds in strawberries but the stability of ester compounds during pressure is still under discussion. HP-treated strawberry purée differed from heattreated and unprocessed strawberry purée. Cross validation of the electronic nose data showed that heat treatment changed volatile compounds more than high-pressure processing. Corresponding results were reported for similarly processed raspberry and black currant purées.

Baxter et al., observed no differences in the concentration of volatile flavour compounds between freshly frozen, heat-treated (85 °C/25 s) or highpressure-treated (600 MPa/18e20 °C/60 s) orange juice. The results of the chemical analysis were supported by the results of a trained sensory panel and a consumer panel, which did not remark any differences in odour or flavor between the differently treated orange juices.

Navarro et al. reported that when HP-treated (400 MPa/ambient temperature/20 min) strawberry purée was stored for 30 days at 4 °C, increases in the contents of methyl butyrate, mesifurane, 2-methylbutyric acid, hexanoic acid, ethyl butyrate, ethyl hexanoate, 1-hexanol and linalool were observed.

The flavour of HP-treated basil (two pulses/860 MPa/75 °C and two pulses/700 MPa/85 °C) was more intense than the aroma of conventionally heatsterilized, frozen or dried basil. A significant volume of important flavour compounds accumulate in many fruits as non-volatile and flavourless glycoconjugates, which are known as glycosidic aroma precursors. These glycosides can be hydrolysed to volatile aglycones by the action of b-glucosidases, enzymes naturally present in many plants.

HP processing is a promising preservation method of fruits and vegetables, even though the original fresh sensory properties are not always fully retained. The sensory properties of many HP-treated fruit and vegetable products are still superior to those of products preserved in the traditional way by heat treatment. Regarding flavour, it is difficult to evaluate how HP-induced changes in volatile compounds affect the overall flavour of fruits and vegetables.

Effect of HPP on the Inactivation of Microorganisms

The basic principles of HPP microbial inactivation are based on protein denaturation which results in enzyme inactivation, and the agglomeration of cellular proteins. Moderate level of pressure (10–50 MPa) decreases the rate of growth and reproduction, whereas higher level of pressure causes inactivation. The cell membrane is constructed as a bi-layer of phospholipids and high pressure causes a phase transition and as a consequence the membrane is destabilized and the permeability is negatively affected. Further, the inactivation could be linked to protein denaturation resulting in the dissolution of membrane bound enzymes. The partial inactivation of enzyme systems by high pressure leads to a breakdown of metabolic actions in biological systems. The protein denaturation depends on such external factors as pH, salt content, water activity (aw) and the presence of other ingredients like sugars. Aside from the product parameters, the processing conditions: pressure (P), temperature (T) and time (t) have a decisive importance on the inactivation of living cells. In the high temperature domain, it is generally accepted that pressure and temperature act synergistically on the inactivation of vegetative bacteria. For the majority of microorganisms, the highest pressure tolerance is found between 20 and 30 °C. In the case that lower temperatures are applied the stability is decreased. Inactivation depends on a number of factors related to the Gram type, physiological state and strain particularities. Synergistic effects with HPP have been described with antimicrobials, low pH, carbon dioxide, vacuum packaging and chilled storage.

Katharina et al., produced a new fermented fish sausage product, based on monkfish. To evaluate food safety, a challenge test was performed, in which the raw materials were inoculated with low levels of Listeria monocytogenes and Salmonella enteric. The product was manufactured, fermented, QDS dried, and half of the samples were pressurized (600 MPa, 5 min, 13 °C). They monitored pathogens, technological microbiota, spoilage indicator bacteria from fish (hydrogen sulphite producing bacteria, coliforms and Escherichia coli) and physicochemical parameters during manufacturing and after 6, 13, 20 and 27 days of refrigerated storage at 4 and 8 °C. Their results showed that in the finished product, pathogens and spoilage indicator bacteria could not grow but decreased and E. coli was not detected during storage. Pressurization had an important reducing effect on technological microbiota, and eliminated L. monocytogenes, S. enterica, hydrogen sulphite producing bacteria and coliforms immediately after production and during refrigerated storage.

Effect of HP Processing on Texture

Due to cell disruption, HP processing facilitates the occurrence of enzymatic and non-enzymatic reactions. Texture changes in fruits and vegetables can be related to transformations in cell wall polymers due to enzymatic and non-enzymatic reactions. Substrates, ions and enzymes which are located in different compartments in the cells can be liberated and interact with each other during HP treatment. At the same time, pressure can enhance the action of pectin methyl esterase (PME), lower the poly galacturonase (PG) activity (occurring mostly at moderate temperature), and retard b elimination [a reaction where loss of two substituents from adjacent atoms (such as carbon, nitrogen, oxygen) results in the formation of new unsaturated bond] (possibly occurred at elevated temperatures). Different pressure and temperature combinations can be used to activate or inactivate some specific pectinases during processing to create textures, which cannot be formed by thermal processing. Moreover, the use of HP processing can be combined with pretreatments such as infusion of exogenous pectinases and/ or soaking in calcium chloride solutions, which can result in increased firmness of the processed fruits and vegetables. HP treatment can disturb the cell permeability of fruits and vegetables, which enables movement of water and metabolites in the cell. The degree of cell disruption is not only dependent on the applied pressure level but also on the type of plant cell. HP processing affects the organization of the parenchyma cells. The plant cells disintegrate and the intercellular spaces are no longer filled with gas (for example in spinach leaf). After HP treatment, cavity formation occurs and a firm texture and a soaked appearance (e.g. cauliflower) are noticed after HP processing.

Concerning HP effects on texture of (solid) fruits and vegetables, hardness or firmness is mostly used as a parameter.

Besides increase in hardness, fruits and vegetables such as apple, pear, orange, pineapple, carrot, celery, green pepper and red pepper experienced softening at pressures above 200 MPa (room temperature/5-60 min). At 100 MPa, pear was the most pressure sensitive fruit followed by apple, pineapple and orange, while at 200 MPa, apple was more sensitive than pear. Softening under pressure was also observed for cherry tomatoes. Pressures from 200 to 400 MPa (20 °C/20 min) resulted in increased texture damage while pressures greater than 400 MPa (500 and 600 MPa/20 °C/20 min) led to less apparent damage. The softening of cherry tomatoes HP treated at 200-400 MPa may be a result of simultaneous activity of PME and PG, since PG is able to depolymerize pectin that has been demethylated by PME. HP treatment can affect the rheological properties of food products such as crushed fruits and vegetables, purée, pulp and juice.

Ahmed et al. reported that the viscosity of mango pulp increased after HP treatments at 100 or 200 MPa (20 °C/15 or 30 min), while a reduction in viscosity was observed after HP treatments at 300 and 400 MPa (20 °C/15 or 30 min).

In the presence of NaCl (0.8%), the effect of pressure was the opposite the viscosity increased with increasing pressure up to 400 MPa. For some fruit juices, cloud stability is an important quality aspect. A shelf-life study on navel orange juice showed that: (i) pressure treatment (600 MPa/40 °C/4 min) resulted in a higher viscosity than thermal treatment (80 °C/60 s) and (ii) a limited cloud loss and a small decrease in the viscosity of HP-treated juice were observed during storage (0, 5, 10, 15 or 30 °C for 64 days) even at an elevated storage temperature (30 °C). It is suggested that residual PME activity is responsible for the quality loss of orange juice during storage.

Effect of HPP on the Texture of Meat and Meat Products

Low pressures (200 MPa) can tenderize pre-rigor meat, whereas tenderization post-rigor with HPP can only be achieved by higher temperatures. In fresh meat, the application of low pressure levels can be used to improve the functional and rheological properties of turkey meat with low pH or PSE meat. The influence of HPP on the meat tenderness is depending on the rigor stage, pressure and temperature level applied, and their combination. Meat tenderization by HPP is likely caused by lysosome breakdownand subsequent proteolytic activity release to the medium. Prerigor treatment of fresh meat by HPP was shown to be very effective to improve the tenderness of fresh meat. However, the application of HPP at pre-rigor state would require the development of hot boning at slaughterhouses. The application of HPP can be used to improve the water retention properties of raw material used for the production of meat products and as a result to the development of products with reduced salt content.

The first pressure treatment of raw material is designed to denature myofibrillar proteins and to create the correct product characteristics of consistency and texture, while the second pressure treatment is carried out after the pressurized raw material is emulsified using raw liver in the bowl chopper to increase the shelf-life and to ensure final product characteristics. Replacing the two thermal steps results in a significantly smoother and homogenous product with an increased liver-taste as well as significant improvements in time and energy consumptions and nutritional value.

Effect of HP Processing on Colour

Chlorophyll is a green compound found in the leaves and green stems of plants. Chlorophylls a and b have different stabilities towards pressure and temperature. At room temperature, chlorophylls a and b exhibit extreme pressure stability but at temperatures higher than 50 °C, HP treatment affects their stability for example, a significant reduction in the chlorophyll content of broccoli juice. HP treatment (at low and moderate temperatures) has a limited effect on pigments (e.g. chlorophyll, carotenoids, anthocyanins, etc.) responsible for the colour of fruits and vegetables. The colour compounds of HP processed fruits and vegetables can, however, change during storage due to incomplete inactivation of enzymes and microorganisms, which can result in undesired chemical reactions (both enzymatic and non-enzymatic) in the food matrix. The pressure dependency of the degradation rate constant of chlorophyll b at 70 °C is higher than that of chlorophyll a. For example, elevating pressure from 200 to 800 MPa accelerates the degradation of chlorophyll a and chlorophyll b of broccoli by 19.4% and 68.4%, respectively. HP treatment at ambient and moderate temperatures results in limited colour change of green vegetables.

In many cases, the green colour of vegetables becomes even more intense (decrease in L*, a* and b* values) for example green beans after HP treatment of 500 MPa/ambient temperature/1 min. This might be caused by cell disruption during HP treatment resulting in the leakage of chlorophyll into the intercellular space yielding a more intense bright green colour on the vegetable surface. However, at elevated temperature, the green colour shifted visibly to olive-green with a concomitant increase in the a* value. During storage, the green colour of the vegetables HP treated at room temperature turned into a pale yellow colour (decrease in a* value) probably due to chemical reactions such as oxidation. By comparison, the vegetables pressurized at elevated temperatures, which results in inactivation of some enzymes, showed no further colour change during storage.

Fengxia et al., used high hydrostatic pressure (HHP, 600 MPa/1 min) and high temperature short time (HTST, 110 °C/8.6 s) treatments of mango nectars were comparatively evaluated by examining their effects on antioxidant activity, antioxidant compounds, color, and browning degree (BD) immediately after treatments and during storage of 16 weeks at 4 and 25 °C. Steam blanching was used prior to HHP and HTST to inactive endogenous enzymes. Their results showed that antioxidant capacity (FRAP assay),

L-ascorbic acid, sodium erythorbate, total phenols, total carotenoids, the redness (a*), the yellowness (b*), and BD changed insignificant after HHP or HTST treatment. The lightness (L*) exhibited a significant decrease in HTST-treated mango nectars, while no significant changes in HHP-treated samples. After 16 weeks storage at 4 and 25 °C, there were significant changes in antioxidant activity, antioxidant compounds, color, and BD of mango nectars, whereas differences between HHP- and HTST-treated samples were not significant except for the decrease in L-ascorbic acid and sodium erythorbate, which was more pronounced in HHPtreated samples.

Carotenoids are important for the orangeeyellow and red appearance of fruits and vegetables. Carotenoids are rather pressure stable. HP treatment increases the extraction yields of carotenes from the plant matrix. Pressure-induced isomerization of all-trans lycopene in hexane was observed during HP treatment at 500 and 600 MPa (room temperature/12 min). This phenomenon was not, however, observed in food matrices such as in tomato puree. The colour of tomato purée remained unchanged after HP treatment (up to 700 MPa) at 65 °C even for 1 h. Anthocyanins are water-soluble vacuolar flavonoid pigments responsible for the red to blue colour of fruits and vegetables. Anthocyanins are stable during HP treatment at moderate temperature, for example, pelargonidin-3- glucoside and pelargonidin-3-rutinoside in red raspberry (Rubus idaeus) and strawberry (Fragaria x ananassa) during HP treatment at 800 MPa (18-22 °C/15 min). Anthocyanins in pressure-treated vegetables and fruits were not stable during storage.

Kyung et al., found that after HHP treatment, the hardness and color values of L* (lightness), a* (redness), and b* (yellowness) of garlic samples decreased, while the cohesiveness value of garlic samples was increased (P < 0.05). Besides the instability of colour pigments, browning plays an important role in the discoloration of HP-treated food products. In fruit-based food products, no visual colour differences (based on L*, a* and b* values) are observed immediately after HP treatments. Ahmed et al. observed that colour parameters such as (a/b), C and h values of mango pulps remained constant after HP treatment indicating pigment stability, while increasing pressure intensity decreased the value of DE. During storage, discoloration of pressurized food products occurred during storage (3 °C) due to enzymatic browning. Colour changes in HP-treated fruits and vegetables can be related to changes in textural properties. This phenomenon was observed in tomato based products. HP treatment (400 MPa/25 °C/15 min) resulted in an increase in the L* value of tomato purée indicating a lightening of the purée surface colour.

Effect of HPP on the color of meat and meat products Studies indicate That HPP provokes drastic changes in freshmeat color, while the changes in curemeat products are acceptable and depending on the water content and aw value. The color of meat depends on the optical properties of the meat surface as well as on the myoglobin content of the muscle. In contrast, the color of cured meat products is mainly created due to the presence of nitrosylmyoglobin, resulting from the reaction of nitric oxide (from sodium nitrite or sodium nitrate) with myoglobin. The effect of HPP on minced beef and concluded that L* color values increased significantly in the pressure range 200–350 MPa, giving themeat a pink color, while a* values decreased at 400–500 MPa, resulting in a gray-brown color.

High pressure caused dramatic changes in the color of fresh meat and thus makes difficult the commercialization of HPP freshmeats since they lack the typical color of freshmeat from the consumer's perspective. By keeping the ratio oxy- to deoxymyoglobin low before pressurization aminor conversion to ferricmyoglobin was observed in a model system.

Studies on cured meat products reported an increase in lightness and a decrease in redness when products are pressurized. But Ferrini et al. showed that the changes in lightness and redness were dependent on the water content of the meat products. HPP treatment increased L* and reduced a* and b* in raw cured hams with high water contents. The HPP treatment had negligible effect on the raw cured hams with low water contents. Bak et al., showed that pork meat treated above 300 MPa became significantly less red and more yellow within the first day of storage. This fact was explained by the formation of a short lived ferrohemochrome myoglobin specie which is transformed into a brown, ferric form of the pigment within the first day of storage.

EFFECTS OF HIGH PRESSURE PROCESS ON FOOD MICROORGANISMS

Potential of High Pressure Process (HPP) in increasing shelf-life of food and improving the microbiological safety of food was reported for the first time more than 100 years ago, and the United States was the first country that studied the process. Hit in 1899 found that HPP can inhibit microorganisms in milk and increase its shelf life, and thus developed HPP in food industry. He tested to see if under different levels of pressure milk stays pleasant compared to unpasteurized milk, and found that using high pressure of 463 MPa for 1 hour, delays milk's getting sourness for at least 24 hours. With the advancement of high pressure equipment that has reached its peak recently, the statistics show that there are about 200 producers of high pressure equipment around the world. High pressure equipment is widely used for production of meat, dairy products, seafood, fruits and vegetables and a variety of beverages. The value food produced with HPP is more than 10 billion dollars and the number of manufacturers increases annually. Evaluation of security of new storage technology depends on a reliable estimate of its performance against pathogenic and food spoiling microorganisms. In the following figure, part A shows the growing trend in the use of HPP in four continents of Oceania, Asia, Europe and America between 1990 and 2008. Pie chart shows the use of this process in the four major food groups, including vegetables, beverages, meat and sea food.

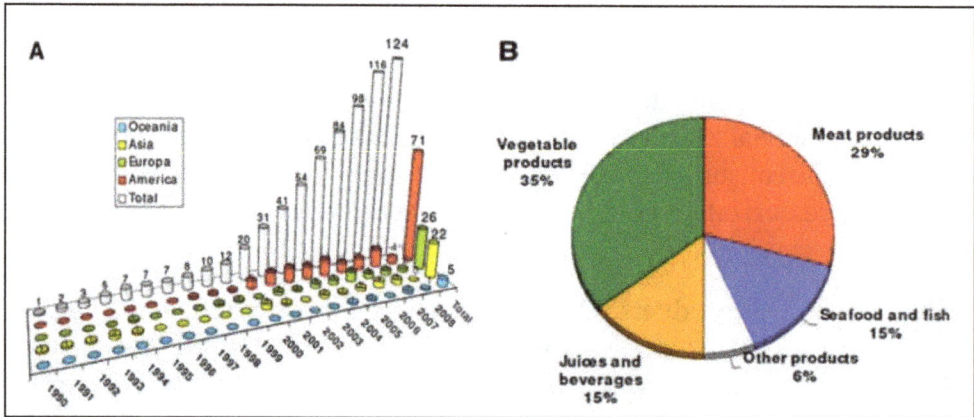

(A) Use of HPP in four continents of Oceania, Asia, Europe and America
(B) Use of HPP in four major food groups (vegetables, beverages, meat and sea food).

For processing in high pressure, high hydrostatic pressure is used. The main country manufacturing high pressure equipment is Japan. Of the main companies manufacturing high-pressure equipment Mitsubishi and Kobe Steel can be named. Equipment producing high hydrostatic pressure exists in many forms. In general, the equipment consists of two parts: process and control. The process part involves under-pressure tank, pump and heating system. Control unit is programmable for controlling the pressure, temperature and time. Needed tool for this process is a mechanical chamber (steel cylinder) and a pump to generate pressures up to several hundred MPa. After that food is placed in a suitable container and capped, the package is placed in a cylinder containing liquid with low compressibility (such as water). Pressure is generated by a pump which may be used permanently (static) or discontinuously. In the last case, two or three cycles of implementing pressure with different times may be used. The main problem with heat sterilization is the loss of a significant amount of taste after processing plants. HPP in fresh basil causes better preserve of the main constituents of essential oil compared to freezing, drying and heat sterilization processes. So, this process preserves the flavor compounds and plant tissue. Nowadays, the vegetables with firmer texture are desirable, and the process better preserve the product color compared to thermal processes. In a way that the color of some products containing chlorophyll are affected during thermal processing while HPP better preserves the color of products such as spinach, carrots and potato puree.

Department of Agriculture of America has considered E. coli O157: H7 as an indicator species for raw materials or finished product and has required a 5-log reduction in time to get to safety and health of food for consumers. Although by using high pressure endospore strain growth cannot be completely prevented, no food processed with HPP at room temperature is supplied in stores now.

Effects of High Pressure on Microorganisms

HPP can be applied to inhibit the growth of microorganisms in food, but there are different types of microorganisms with different physiological features, and many of them may have different characteristics of resistance to pressure. Table shows deactivating microorganisms in meat by HPP.

Table: Inactivating microorganisms in meat by HPP.

Target microorganism	Product	Initial counts (Log 10 cfu/g)	Cell load reduction (Log 10 cfu/g)	Treatment
L. monocytogenes	Cooked ham 2.6	2.6	1.9, after 42 days at 6 °C	400 MPa, 10 mm
Total viable count	Bovine meat	4.0	2.5, after treatment	520 MPa, 4.2 min, 10 °C
L. monocytogenes	Dry-cured ham	4.65	Total inactivation (<1 cfu/g)	600 MPa, 9 min
S. aureus	Marinated beef Dry-cured ham	362	2.67, after treatment	600 MPa, 6 min, 31 °C
Satmonetia spp.	Cooked ham	3.8	Total inactivation (<1 cfu/g)	600 MPa, 6 min, 31 °C
E. coli 0157:H7	Raw minced meat	5.9	5, after treatment	700 MPa, 1 min, 15 °C
L monocytogenes	Sliced beef cured ham	4.0	2, after 210 days, 6 °C	500 MPa, 5 min, 18 °C
C. freundii	Minced beef muscle	7	5, after treatment, 20 °C	300 MPa, 10 min, 20 °C
L. monocytogene	Iberian ham	6.3	3.6, after 60 days, 8 °C	450 MPa, 10 min, 12 °C
Lactic Acid Bacteria	Marinated beef	4.94	3.99, after treatment 6	600 MPa, 6 min, 31 °C
Campylobacter jejuni	Pork slurry	6-7	6, after treatment	400 MPa, 25 mm., 10 °C
Toxoplasma gondii	Ground pork meat	Viable tissue cysts	Inactivation (cysts not viabtes)	300 MPa

Generally, increase of pressure causes different levels of impact on microorganisms. Table shows how a commercial sterilization process can be accomplished through a two pulse process (pulse length 1 minute and interval 30 seconds). This table shows that pressure increase can be reduced by the initial temperature of the product.

Table: Two-pulse high pressure sterilization process for some food.

Type of Food	Initial temperature (°C)	Pressure (MPa)
Most vegetables, pasta, meat	90	690
Most vegetables, potatoes	80	828
Vegetables, seafood, potatoes	70	1172
Eggs, milk	60	1700

According to the studies, 50 MPa pressure can inhibit protein synthesis in microorganisms and reduce the number of ribosomes. Pressure of 100 MPa could partly lead to denaturation of proteins and 200 MPa pressure causes damage to the cell membrane and the internal structure of the cell. Rising the pressure to 300 MPa or more causes irreversible denaturation of the enzymes and proteins, resulting in rupture of cell membranes and withdrawal of intracellular contents which will lead to bacteria death. Thus, the effects of high pressure on microorganisms can be primarily divided to changes in cell morphology, inhibition of metabolic reactions essential for cell survival, and changes in the genetic mechanism.

Morphology

When microorganisms are under the influence of high pressure, the cell membrane is the usually the first organ that suffers. The main function of the cell membrane is to maintain cell morphology, balance the pressure inside and outside the cell and regulate the entry and exit of foreign substances from outside to cell. When high pressure damages cell membrane and its structure, the absorption of nutrients in the cell of microorganisms is affected, the disposal of waste products in the cells is disrupted and the natural pathway of metabolism is destroyed.

It seems that bacterial cell membrane is one of the objectives of HPP. Increase of pressure around the cells causes swelling and disruption of membrane structure and disrupts the permeability of the membrane, which eventually leads to cell death. Disruption of membrane structure is the result of changes in morphology and physical properties of the cell.

Changes in the Genetic Content

HPP can have negative effects in genetic factors in microorganisms such as DNA replication and gene expression. Studies show that HPP inhibits DNA replication and gene expression enzymes. Studies have found that some microorganisms have regulating genes to adapt to their environment (such as E. coli O157:H7, gram-positive bacteria such as Bacillus subtilis, Staphylococcus aureus and Listeria monocytogenes) so that when placed in an unsuitable environment for growth, stress-resistant strains of the bacteria appear.

Factors Affecting the Pressure-resistance Properties of Microorganisms

Most of microorganisms have great ability to adapt efficiently to the environment. In unsuitable conditions, microorganisms use different protection mechanisms such as compliance with environmental, changing to dormant state (endospore), activation of expression of genes resistant to stress or production of resistant mutants. Thus, stress tolerance in microorganisms is not fixed, but they are influenced by internal and external factors, such as type of microorganisms, stage of development and the environment.

Diversity of Micro-organisms

Comparing the studies carried out show that prokaryotes resistance to pressure is more than that of eukaryotes, Gram-positive bacteria is more than that of Gram-negative, and cocci bacteria are more resistant than Bacillus. Smaller and cocci-shaped bacteria are generally more resistant to HPP than larger rod-shaped bacteria. In Table the resistance of different microorganisms with D value is displayed during high pressure. In difficult environments for growth, some microorganisms to escape from stress or harsh conditions form endospore. Under such conditions, the resistance to morphological pressure of endospore is substantially increased compared with the original growth of microorganisms.

Table: The difference between D-value of several different microorganisms in HPP.

Microorganisms	Pressure (MPa)	Temperature (°C)	D-value (min)
Clostridium pasteurianum	700	60	2.4
Citrobacter freundii	230	20	14.7
Listeria monocytogenes	414	25	2.17
Salmonella typhimurium	414	2	1.48
Listeria innocua	400	2	3.12
Staphylococcus aureus ATCC 27690	250	20	15

Growth Phase

In general, microorganisms are more resistant to pressure in stationary phase compared to exponential phase of growth. It is for the reason that during the growth phase, microorganisms are continuously involved in duplex cell division and synthesis and their stress tolerance reduces in difficult conditions. Microorganisms in the stationary phase have a full structure and are under membrane protection, so they can tolerate higher levels of stress.

Environmental Circumstances

According to Herdel's technology concept, when high pressure is used for the pasteurization of food, a combination of poor environmental conditions for the growth of microorganisms such as low pH and temperatures higher or lower than the suitable growth temperature reduce the resistance of microorganisms to pressure and facilitate its destruction. Combined use of these methods, is more efficient than using them separately in deterring microorganism growth, this means that combination of maintaining factors can dramatically increase food quality. For high-pressure pasteurization process, the combination of high pressure and temperature are considered as the most appropriate combination.

HPP has recently attracted the attention of researchers for different studies. The most successful commercial and industrial applications of HPP are to deactivate microorganism growth, to achieve pasteurization without heat, to increase shelf-life of food products without changing nutritional, functional and sensory properties of food. In traditional thermal and conventional food processing techniques, the main role of temperature is to inhibit the growth of pathogens and to ensure food security. Thus, the value of pasteurization at different temperatures should be carefully evaluated. High pressure can inhibit the growth of pathogens and maintain the quality of the food and just keep it the same as normal food with minimal processing. Although high pressure processing has developed for more than 100 years in America and Japan, in many countries it is still a new emerging trend. All new methods of processing should comply with food safety standards before commercial use.

EFFECTS OF HIGH PRESSURE PROCESSING ON BIOAVALIABILITY OF FOOD COMPONENTS

Foods provide essential nutrients, growth factor, immune boosting compounds and other bioactive compounds (vitamin C, carotenoids, phenolic compounds, vitamin E, glucosinolates) with antioxidant, antitumoral and anticancerogenic properties for human body. Today, the link between bioactive compounds taken by the diet and occurrence of some disease are well known. As lack of some nutrients causes some health problems such as scurvy, beri beri, etc., compounds formed by food processing technologies such as hydroxy methyl furfural (HMF), acrylamide, aldehyde and ketones can cause some metabolic disorders and even cancer. The diet is also associated with the morbidity and mortality in the chronic diseases, such as cardiovascular disease, cancer, hypertension and obesity. It is reported by several studies that diet is attributed to one-third of all cancer cases and one-half of cardiovascular diseases and hypertension.

Thus, it is important that consumed food and food compounds should provide essential nutrients for human body with minimization of health hazard effect. On the other hand, the metabolic pathway of each compound in addition to their bioavailability is important to determine their usage by human body. In order to be bioavailable, a food compound must be released from the food matrix and modified in the gastrointestinal (GI) tract. Moreover, the stability of the compound affecting its bioavailability and their possible beneficial effects is important before concluding on any potential health effect.

Bioaccesibility and its Measurement

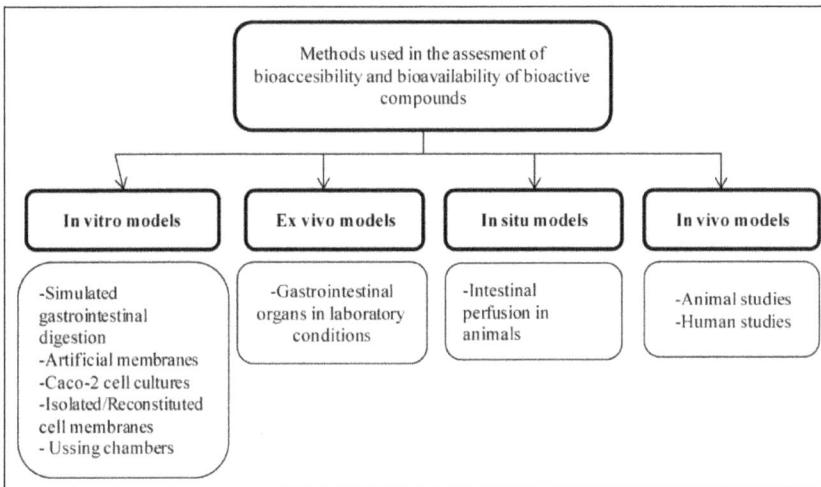

Bioavailability.

One of the important terms to explain the use of a bioactive compound by human body, bioaccessibility, is defined as the fraction or the quantity of the compound released from the food matrix in GI tract which can be absorbed. Food need to go through digestive transformations to be converted into material ready for assimilation, the absorption/assimilation into intestinal epithelium cells, and lastly, the pre systemic metabolism (both intestinal and hepatic) need to occur for bioaccesibility of a component. If the definition is solely on absorption based, then the beneficial effects of unabsorbed nutrients (such as binding of bile salts by calcium in the tract) would be missed. Therefore, generally simulating gastric and small intestinal digestion, sometimes followed by $CaCo_2$ cells uptake is the usual evaluation for in vitro digestion procedures for some nutrients. Utilization of a specific nutrient is also important, and thus, the term bioavailability is also described as the utilization of a nutrient, and therefore, can be defined as the fraction of ingested nutrient or compound that reaches the systemic circulation and is utilized.

In general, bioavailability involves GI digestion, absorption, metabolism, tissue distribution, and bioactivity of a certain compound. Studies including bioavailability of a compound –therefore- must reveal that the component analyzed is efficiently digested and assimilated and then, once absorbed, exerts a positive effect in human health with considering bioactivity. Bioavailability of a compound need to be determined in vivo in animals or humans as the area under the curve (plasma-concentration) of the compound obtained after administration of an acute or chronic dose of an isolated compound or a compound-containing food; whereas bioactivity is the specific effect upon exposure to a substance including tissue uptake and the consequent physiological response (such as antioxidant, anti-inflammatory) that can be evaluated in vivo, ex vivo, and in vitro).

Transformation of the fraction of food components by digestion into potentially accessible matter through all physical–chemical processes that take place in the lumen is defined as digestibility. Assimilation is another term that needs to be defined, and it refers to the uptake of bioaccessible material through the epithelium by some mechanism of transepithelial absorption.

Food processing technologies may alter, change or diminish the bioaccessibility of a specific compound by denaturation or causing changes in the structure. For most of the compounds, changes in the physical structure or denaturation may even direct toxic, mutagenic, carcinogenic and teratogenic effects. It is reported by numerous studies that consumers have a growing preference for convenient, fresh like, healthy, minimal-processed food products with natural flavor and taste and extended shelf-life. In order to meet these demands, alternative not-thermal preservation technologies as high pressure processing (HPP), pulsed electric fields (PEF) irradiation, light pulses, and natural bio-preservatives together with active packaging without compromising safety have been proposed. HPP carried out with intense pressure in the range of 100-1000 MPa with or without heat is a promising "non-thermal" technique for food preservation allowing most foods to be preserved with minimal effect on taste, texture or nutritional characteristics.

HPP was found superior to thermal sterilization and pasteurization due to the maintenance of sensory and nutritional characteristic of treated food products. Both liquid and high-moisture-content solid foods are subjected to pressure treatment at relatively low temperature being lethal to microorganisms, but not effective to covalent bonds representing a unique characteristic of this technology because HPP has a minimal effect on food chemistry. HPP retains food quality while avoiding the need for excessive thermal treatments or chemical preservation; however positive effect of HPP on bioavailability and bioaccessibility of food components need to be proven in order to claim that this technology provides preservation of functional properties. It is well known fact exposure of plant foods to HPP having changes at different magnitude on the nutritional properties and possible protective effects of the food once processed causes alteration on plant matrix structures. For example, HPP processing of green beans at 600 MPa provides a significant increase in lutein availability compared to untreated samples and this positive effect can be due to facilitated release of lutein within the plant tissue matrix during the in vitro digestion process by the disruption of cellular structures in the beans by exposure to high pressures. It is recommended by studies that possible changes on plant the tissue matrix such as disruption of plant cell walls induced by HPP results in the release of compounds with antioxidant actions into the extracellular environment.

Even though bioaccessibility and bioavailability of polyphenols, carotenoids, and glucosinolates may be enhanced by HHP as their extractability is increased by HPP, the extractability and bioavailability of nutrients and correlation between process-induced matrix disruption are not directly related to each other. Bioassesibility of certain compounds, carotenoids for example, is more complex, in that, the positive effect of HPP would be anticipated based on its effect on food matrix structure. Moreover, it is revealed that the impact of processing on bioaccessibility and bioavailability of nutrients. In general, is dependent on the type of nutrient, the structure and composition of the food matrix and the processing technique employed.

Although bioavailability or bioaccessibility of different food components were reported by both in vivo and in vitro studies, the effects of the food matrix on the bioavailability or bioaccessibility of antioxidant minerals and starch as well as effect of HPP have not been reported. Digestion and absorption of starch can be influenced by the direct interactions between this component and some components of food, such as binding to proteins and polysaccharides. Although HPP was proven to preserve nutritional properties; processing of apples at 500 MPa for 2, 4, 8 and 10 min significantly affect the antioxidant activity, mineral and starch content and bioaccessibility of apple samples. In vitro digestion has a noticeable effect on the antioxidant concentration, half maximal inhibitory concentration (IC_{50}), with much lower values (a smaller IC_{50} value corresponds to a higher antioxidant activity) of apple samples compared with those untreated and nondigested. Calcium, iron and zinc bioaccessibility of apple samples in vitro was calculated as the percentage of the element dialyzed of the total amount present in the aliquot (% dialyzability) by the following equation,

$$\text{Dialysis}(\%) = 100 x \frac{Y}{Z}$$

where Y is the element content of the dialysates mineral fraction (mg mineral element/100 g), and Z is the total mineral (calcium, iron or zinc) content of the sample (mg mineral element/100 g grain). Calcium, iron and zinc content of unprocessed samples are measured as 30.33 ± 1.94 mg/100 g, 14.46 ± 3.49 mg/100 g and 6.22 ± 0.91 mg/100, respectively and HPP provides an increase in the mineral content availability by 2.11-303.00% for calcium, 4.63-10.93% for iron and 8.68-28.93% for zinc. Moreover, both the dialysability and solubility of calcium, iron and zinc with respect to the values for the untreated sample are found to be reduced by HPP. HPP-treated samples exhibit higher antioxidant capacity with in vitro digestion and long time than that of the non-digestion samples suggesting an increase in the amount of antioxidants released by the apple matrix into the human intestine, and hence the antioxidant capacity of these samples, may be higher than expected from the data based on chemical extracts. Antioxidants are potentially available in the small gut; the degree to which they produce an antioxidant effect depends on the rate of absorption and this fact need to be considered when evaluating the antioxidant capacity of a fruit from a nutritional standpoint. Moreover, if the antioxidants are not released in the digestive enzymatic extracts, they may enter the colon where they can be fermented by the microflora, yielding different compounds that may be metabolized and may provide an antioxidant environment. Consumption of apple under HPP may supply substantial antioxidants, mineral and starch, which may provide health promoting and disease preventing effects.

The sum of starch and the product of starch degradation not absorbed in the small intestine but is fermented in the large intestine of healthy individuals is described as resistant starch (RS). It has been released that RS participates in the reduction of glycemic and insulinemic responses to food and it has hypocholesterolemic effects and protects against colon rectal cancer. Thus, the RS content in food, as well as the digestive rate and level of starch, has a positive effect on health. It is indicated that increased treatment time at 500 MPa from 2 to 10 min causes significant decrease in RS content of the apple samples compare to untreated ones having 19.6 ± 1.9% of RS. Moreover, total starch (TS) of the untreated samples (81.2%) have increased to 99.8 % under 500 MPa for 10 min processing conditions. The TS is measured as = digestible starch (DS) + RS, and it ranges between 95.1% (untreated sample) and 103.5% (500 MPa for 10 min). Based on the HPP processing of apple, it can be said that consumption of HPP may supply substantial antioxidants, minerals and starch which may provide health promoting and disease preventing effects.

Effect of HPP at 0.1, 100, 300 and 500 MPa for 10 min on mineral elements, amino acids (AAs), antioxidants and starch on germinated brown rice (BR) at 37°C for 36 h has revealed that while the in vitro bioaccessibility of calcium and copper increases by 12.59-52.17% and 2.87-23.06% after HHP, respectively, bioaccessible iron is decreased. Effect of HPP is obvious especially on AAs in that particularly indispensable AAs and gama-aminobutyric acid, as well as bioaccessible total antioxidant activities and starch resistance to enzymatic hydrolysis, significantly improved by HHP. Starch digestibility, on the other hand, increased by germination. HPP above 300 MPa causes structural changes in bran fraction revealing relationship between germination and HPP on nutrients bioaccessibility and develop appropriate processing conditions.

Changes on bioaccessibiliy of carrot, broccoli and green beans carotenoids is investigated after HPP processing of 400–600 MPa pressure and 2 min processing time. Among those, the bioaccessibility of carotenoids in carrot is not significantly changed. However, a slight improvement in lutein bioaccessibility in green beans and reduction on β-carotene bioaccessibility in broccoli is observed by HPP at 600 MPa.

Bioaccessibility of some compounds can be affected by physical state of the plant tissue. For example, the bioaccessibility of β-carotene in carrot tissue is inversely related to hardness. Both mild thermal pasteurization and mild HPP pasteurization cause similar reduction in firmness, but tissue softening caused by these two processes have different physical effects on cell structure. Tissue softening provided by mild HPP in carrot tissue is mainly due to turgor pressure loss explaining the higher bioaccessibility of β-carotene in the sample compared to mild thermal pasteurization. On the other hand, thermal processing applied at more intense pasteurization and sterilization conditions results in a softer tissue compared to HPP resulting in higher bioaccessibility.

Studies related to bioavailability of high pressure processed orange juice vitamin C is somewhat controversy. Although no significant difference is observed between the fresh and HP processed orange juice vitamin C in one study, consumption of 500 mL per day of HP-treated orange juice at 400 MPa pressure, 40 °C processing temperature for 1 min significantly elevates plasma vitamin C concentration in healthy subjects, indicating that vitamin C in HP-treated orange juice is bioavailable. In addition, the level of biomarkers of oxidative stress and inflammation, F2-isoprostanes, uric acid, C-reactive protein and prostaglandin E2 also are lowered significantly at the end of the 14-day study, indicating that consumption of orange juice may help to decrease the risk of chronic diseases. HP- treated 'gazpacho' a traditional Spanish vegetable soup shows similar effect as the orange juice treated at 400 MPa pressure, 40 °C processing temperature for 1 min.

Effect of HPP on antioxidant capacity, mineral and starch bioaccessibility of a nonconventional food: "algarrobo" Prosopis chilensis seed pressurized at 500 MPa during 2, 4, 8 and 10 min is measured through the antioxidant activity, mineral and starch content and bioaccessibility. All treatments provide an increase in the bioaccessibility of the antioxidant activity (IC_{50}), minerals (dialysis and solubility) and starch (resistant and digestible) compared to the untreated sample of algarrobo samples. Bioaccessibility of calcium, iron and zinc in the treated sample for 500 MPa at 10 min, expressed as percentage solubility, is found several-fold higher (three, three and five times, respectively) than that of the untreated sample. Similar effect is also observed in IC_{50} value in that the untreated samples exhibit the lowest antioxidant activity (0.11 ± 0.005 mg/ mL) followed by all treated samples at 500 MPa for 2, 4, 8 and 10 min.

HPP may also influence the bioavailability of nutrients through its effect on the activity of endogenous enzymes such as folates in vegetable sources usually exist in the less bioavailable polyglutamate forms. The endogenous enzyme, γ-glutamyl hydrolase, hydrolyses the polyglutamyl folates into the more bioavailable short-chain or monglutamyl folates during cell decompartmentalization. Once this enzyme is inactivated by different processes such as freezing usually preceded by blanching, keeping the long chain polyglutamyl folates causes reduction of bioavailability of folates. Unlike freezing usually preceded by blanching, cell decompartmentalization without inactivation of the enzyme and facilitation of the hydrolysis of long chain polyglutamyl folates potentially improving their bioavailability is enabled by HP treatment.

Effect of different processing technologies at both pilot and large scales and comparison of the pilot scale study with blanching (boiling water, 10 min); freezing (−18 and −80 °C) followed by refrigerated (4 °C) thawing and blanching; freeze-drying, followed by rehydration, 6 h refrigeration and blanching; and HP (50–200 MPa/RT/5 min), followed by 6 h storage and blanching on the formation of monoglutamate folates in leeks results in highest yield of monoglutamate folates when the samples are processed by HP (200 MPa/RT/5 min) and freezing (−18 °C/16 h) and thawing (4 °C, 24 h). Among the larger scale process of freezing–thawing–blanching and HP (200 MPa/RT/5 min), blanching causes the highest loss of total folates of 85% in leeks, 65 and 55% respectively in cauliflower and 79 and 81%, respectively in green beans. Blanched or steamed samples regardless of the subsequent process provide the lowest total losses of 16- 38% in the samples. Both blanching and steaming cause a reduction in the proportion of monoglutamate folates in all the vegetables compared to the raw sample from 33% to 6 and 11% respectively in leeks, from 9% to 3 and 2%, respectively in cauliflower and from 33% to 7 and 2%, respectively in green beans. The samples subjected to blanching prior to freezing or HP present similar reductions. However, samples subjected to HP or HP followed by blanching results in an increase in proportion of monoglutamate folates to 74 and 65%, respectively in leeks and 82 and 72%, respectively in green beans, whereas the effect is less significant in cauliflower with 9 and 12%, respectively after HP and HP–blanching. As a result, it is seen that freezing–thawing–blanching has a similar effect of significantly boosting the proportion of monoglutamate folates in all the vegetable.

HPP at 300, 450 and 600 MPa and 30 °C for 0 and 5 min to quantify the influence of pressure on the polyglutamyl chain of 5-methyltetrahydrofolate (5MTHF) in carrot greens, baby carrots and cauliflower provides a significant increase in the conversion of 5MTHF to short chain and monoglutamyl forms and the highest extent of conversion to the monoglutamyl forms at 600 MPa and 5 min with 4, 23, and 2.5 fold increases of the monoglutamyl content in cauliflower, baby carrot, and carrot greens, respectively.

References

- Schaschke, Carl (2010). Developments in High Pressure Food Processing. New York: Nova Science Publishers, Inc. P. 5. ISBN 978-1-61761-706-5

- Couzin, J. (2002). "Weight of the world on microbes' shoulders". Science. 295(5559): 1444–1445. Doi:10.1126/science.295.5559.1444b. PMID 11859165

- Balasubramaniam, V.M., Barbosa-Cánovas, Gustavo V., Lelieveld, Huub L.M (2016). High Pressure Processing of Food Principles, Technology and Applications. Springer. ISBN 978-1-4939-3234-4

- Effects-of-high-pressure-processing-on-bioavaliability-of-food-components: longdom.org

- "High-tech process "shucks" Maine lobster, competes with Canadians". Workingwaterfront.com. Retrieved 2014-03-19

Food Freezing

Food freezing is one of the methods of food preservation which inhibits microbial growth by lowering the temperature. It turns residual moisture into ice, thus inhibiting the growth of most bacterial species. This chapter closely examines the key concepts of food freezing such as thermal properties of frozen food and industrial freeze drying to provide an extensive understanding of the subject.

Freezing, in food processing is a method of preserving food by lowering the temperature to inhibit microorganism growth. The method has been used for centuries in cold regions, and a patent was issued in Britain as early as 1842 for freezing food by immersion in an ice and salt brine. It was not, however, until the advent of mechanical refrigeration that the process became widely applicable commercially. In 1880 a cargo of meat shipped from Australia to Britain under refrigeration accidentally froze, with such good results that the process was at once adopted for long-distance shipments and other storage. In the 20th century quick, or flash, freezing was found to be especially effective with certain types of food.

Except for beef and venison, which benefit from an aging process, meat is frozen as promptly as possible after slaughter, with best results at temperatures of 0 °F (−18 °C) or lower. Fruits are frozen in a syrup or dry sugar pack to exclude air and prevent both oxidation and desiccation.

Most commercial freezing is done either in cold air kept in motion by fans (blast freezing) or by placing the foodstuffs in packages or metal trays on refrigerated surfaces (contact freezing).

THERMAL PROPERTIES OF FROZEN FOOD

The food product properties of interest when considering the freezing process include density, specific heat, thermal conductivity, enthalpy, and latent heat. These properties must be considered in the estimation of the refrigeration capacity for the freezing system and the computation of freezing times needed to assure adequate residence times. The approach to prediction of property magnitudes during the freezing process depends directly on the relationship between unfrozen water fraction and temperature.

It is important to study thermal properties of foods because they affect the design of food processing equipment. The food products undergo changes in composition during such process as freezing, evaporation and dehydration. There are different methods available to measure the thermal properties of food, but the available data differ depending on the method used. The important thermal properties of food are as follows:

Specific Heat

A measure of the heat required to raise the temperature of a substance. When the heat ΔQ is added to a body of mass m, raising its temperature by ΔT, the ratio C given in Equation is defined as the heat capacity of the body.

$$C_p = \Delta Q / \Delta T$$

The specific heat capacity of a food product can be predicted, based on product composition and the specific heat capacity of individual product components. The following expression was proposed,

$$Cp = \Sigma \, (C \, psi. \, m \, si \,)$$

where each factor on the right-side of the equation is the product of the mass fraction of a product component and the specific heat capacity of that component. The specific heat values for product components were estimated by Choi and Okos. The above equation can be used to predict the specific heat capacity of product solids by removing the term for the water fraction. These specific heat magnitudes for the product solids can be used in the prediction of product enthalpy and apparent specific heat.

$$C_p = 4.180 \, X_w + 1.711 \, X_p + 1.98 \, X_f + 1.547 \, X_c + 0/908 \, X_{a,} \, kJ/kg^0C$$

where, X_w: Water fraction,

   Xp: Protein fraction,

   X_f: Fat fraction,

   X_c: Carbohydrate fraction,

   X_A: Ash fraction.

Thermal Conductivity

Thermal conductivity (λ) is the intrinsic property of a material which relates its ability to conduct heat. Heat transfer by conduction involves transfer of energy within a material without any motion of the material as a whole. Conduction takes place when a temperature gradient exists in a solid (or stationary fluid) medium. Conductive heat flow occurs in the direction of decreasing temperature because higher temperature equates to higher molecular energy or more molecular movement. Energy is transferred from the more energetic to the less energetic molecules when neighboring molecules collide.

Thermal conductivity is defined as the quantity of heat (Q) transmitted through a unit thickness (L) in a direction normal to a surface of unit area (A) due to a unit temperature gradient (Δ T) under steady state conditions and when the heat transfer is dependent only on the temperature gradient. In equation form this becomes the following:

$$\text{Thermal Conductivity} = \text{heat} \times \text{distance} / (\text{area} \times \text{temperature gradient})$$

$$\lambda = Q \times L / (A \times \Delta T)$$

The thermal conductivity magnitudes of most food products are a function of water content and the physical structure of the product. Many models suggested for prediction of thermal conductivity are based on moisture content and do not consider structural orientation. The Choi's and Oko's Model for prediction of thermal conductivity is as follows,

$$K = 0.58 \, X_w + 0.155 \, X_p + 0.25 \, X_c + 0.16 \, X_f + 0.135 \, X_a, \, W / m \, ^\circ K$$

where, X_w : Water fraction,

X_p : Protein fraction,

X_f : Fat fraction,

X_c : Carbohydrate fraction,

X_a : Ash fraction.

Thermal Diffusivity

A measure of the rate at which a temperature disturbance at one point in a body travels to another point. It is expressed by the relationship K / dC_p, where K is the coefficient of thermal conductivity, d is the density, and C_p is the specific heat at constant pressure. Very little thermal diffusivity data are available, but it can be determined using relationship of specific heat, thermal conductivity and mass density of the food product.

Freezing Point Depresssion

Probably one of the more reveling properties of water in food is the freezing point depression. Since all food products contain relatively large amounts of moisture or water in which various solutes are present, the actual or initial freezing point of water in the product will be depressed to some level below that expected for pure water.

The magnitude of this freezing point depression becomes a direct function of the molecular weight and concentration of the solute in the food product and in solution in the water.

Thermodynamics of Food Freezing

Freezing is one the more common processes for the preservation of foods. It is well known that lowering the temp reduces the activity of microorganisms and enzyme systems, thus preventing deterioration of the food products. In addition to the influence of temp reduction on m.o. and enzymes, crystallization of the water in the product tends to reduce the amount of liquid water in the system and inhibit microbial growth or enzyme activity in the secondary action.

The engineering aspects of food freezing include several interesting areas. In order to design a refrigeration system that will serve a food freezing process, some indication of the refrigeration requirements or enthalpy change which occurs during product freezing is required. This aspect is related to the type of product being frozen. The second aspect of food freezing that is closely related to engineering is the rate at which freezing progresses. This aspect is related to the refrigeration requirement, but the temperature difference existing between the product and freezing medium are also of significance. The rate of freezing is also closely related to product properties and quality. Product properties resulting from very rapid freezing are significantly different from those obtained by slow freezing. This difference is dependent primarily on the manner in which ice is formed within the product structure. In addition, the rate of freezing will establish the rate of production for a particular food-freezing operation. For this purpose the most rapid rate of freezing is desirable provided that product quality is not sacrifice.

Example:

A formulated food product contains the following components – water 80%, protein 2%, carbohydrate 17%, fat 0.1% and ash 0.9%. Predict the specific heat in W/kg K using Choi's and Oko's model.

Solution:

$$C_p = 4.180\ X_w + 1.711\ X_p + 1.98\ X_f + 1.547\ X_c + 0/908\ X_a$$

$$= 4.180\ (0.8) + 1.711\ (0.02) + 1.98\ (0.001) + 1.547\ (0.17) + 0.908\ (0.009)$$

$$= 3.651\ kJ\ /\ kg°C$$

$$= 0.8726\ kCal\ /\ kg°C$$

$$= 1.0147\ W\ /\ kg°C$$

Example:

Calculate the thermal conductivity of milk using Choi and okos model, if milk contains 87.5% water, 3.7% protein, 3.7% fat, 4.6% lactose and 0.5% ash at 10 °C.

Solution:

$$K = 0.58\ X_w + 0.155\ X_p + 0.25\ X_c + 0.16\ X_f + 0.135\ X_a$$
$$= 0.58\ (0.875) + 0.155\ (0.037) + 0.25\ (0.046) + 0.16\ (0.037) + 0.135\ (0.005)$$
$$= 0.49 + 0.005735 + 0.0115 + 0.00592 + 0.000675$$
$$= 0.51383\ W\ /\ m\ °K$$

FOOD FREEZING SYSTEMS

There are a number of freezing methods and equipment were developed for speed, quality, or for specific types of food. The conditions surrounding the product during freezing are maintained by an enclosure and a refrigeration system.

Freezing systems can be classified in two groups:

- Direct Contact Systems.

- Indirect Contact Systems.

Direct Contact Systems

In direct contact system there is a direct contact between the food product and medium used for reduction of product temperature is used.

The most important types of direct freezing systems are:

- Air blast freezing system.

- Fluidized-bed freezing system.

- Immersion freezer.

Indirect Contact Systems

There are numerous systems used to freeze food products without direct contact between the product and the medium used for the product temperature. Most frozen food are the result of using indirect contact types of freezing systems, where food is separated from the refrigerant by some barrier.

The important types of indirect freezing systems are:

Plate Freezers

It is an indirect freezing system in which the food product does not have any direct contact with the medium used for the product temperature. The basic system of plate freezers consists of flat hollow plates, refrigeration coil these plates to cool the surface in contact with the food products. The plates are made up of aluminum or mild steel.

Aluminum plates are formed from a number of hollow, extruded sections, butt-welded together along their length. The sections are connected at each end to header, allowing ammonia to flow to and fro, each section of the plate. The food products are placed between stacked parallel plates and then pressure applied to the overall stacks to minimize the thermal contact resistance between the plates and product.

The product should be in planar geometry, so unpackaged meat and fish products are suited for this method. Other irregularly shaped products including vegetables such as cauliflower, spinach, broccoli and shrimp can be frozen using this method by packaging the product in brick shaped container prior to freezing. The refrigerated temperature of this method is around -30 °C and it takes less than 24 hours to freeze cartons of chilled and boned meat to below -10 °C. There are two types of plate freezer; horizontal plate freezers and vertical plate freezers.

Horizontal Plate Freezers

A horizontal plate freezer consists of a set of parallel, refrigerated plates within an insulated enclosure. The stack of plates is placed inside a steel frame and each plate connected to the adjacent plate allowing the plates to be moved apart to form a gap for cartons to be placed between two adjacent plates.

Hydraulic rams are used to move the plate and to apply the pressure on the cartons between the plates during freezing. The freezers are operated either batch or continuous modes. In the batch mode, the spacing in the plates is expanded to allow the product to be loaded on the large tray.

Once the products loaded, the plates are hydraulically closed between the refrigerated plate and the food product. Once the freezing process is over, the plates are opened and the product is removed. In the continuous mode the plates are moved through the enclosed system as the product freezes and then the product between the plates removed one at a time.

Vertical Plate Freezers

It is primarily used to freeze unpacked food such as fish. Here the food product is placed directly between the plates and then pressure is applied to assure proper thermal contact. Once the process has been completed, the plates can be opened to remove the product.

Advantages of Plate Freezing Methods

The main advantages of plate freezing method are mentioned below:

- Faster temperature reduction and freezing time particularly suited to hot offal and hot boned meat.

- Better stowage density in container.

- Lower refrigeration capacity requirements due to the absence of high capacity fans in the freezer.

- High quality of the products.

- Speed of freezing is high.

- Significant reductions in energy, packaging and resource consumption.

Disadvantages of Plate freezing Methods

Main disadvantages of plate freezing methods are:

- Higher capital cost.

- A larger charge of refrigerant is required.

- All cartons should be the same height to simplify loading.

Air Blast Freezers

Air blast freezers are used for both direct and indirect freezing applications. In this method, the product is exposed to a low temperature and high velocity air steam. Air blast freezers typically are operated at temperatures of -30 to -45 °C with the air velocities of 10-15 m/sec.

Air-blast freezers have package that acts as the barrier, for product with unusual shapes, short freezing times are possible by maintaining high air velocities, low air temperature, good contact between the package and the product surface. This process can be continuous or batch, most systems are continuous.

In this method a refrigeration system first cools an air streams to a minimum of -40 °C. The air flows through an enclosed system in which the food product is moved through the system on a conveyor belt. The configuration of the conveyor belt system can vary depending on the food products.

In the simplest design of conveyor belt is that the product is conveyed along a straight belt through on tunnel freezer. But in some other design includes multi-tiered and spiral conveyor belts which are used to minimize the space requirements of the system.

The most important factor of the air blast freezer is to ensure uniform air flow over the product.

The direction of the air flow is also an important factor. The best method is parallel air flow (the air flows parallel to the product in the same direction). The other methods of the air flows are counter flow (air flows in the opposite direction to product) and cross flow (air flows perpendicular to the product).

The counter flow methods are effective methods than the other two. The batch air blast freezers are usually consists of a well-insulated container with an air cooler and fans. The product is loaded on a movable belt with stacked trays and the belt is moved into the freezing container. Maintaining uniform air flow is very important, so suitable facilities are required to ensure uniform airflow over all the trays for proper freezing.

Immersion Freezers

This is another important method for freezing which uses direct contact systems. Here the freezing is done with the help of refrigerant. In this method the refrigerant is come in contact with unpackaged food. So this refrigerant must be non-toxic, pure, clean, free from foreign taste, odour, color or bleaching agents and so on.

The refrigerants are used for the immersion freezing are classified in to two classes:

- Low freezing point liquids – These are chilled by indirect contact with another refrigerant.

- Cryogenic liquids – Compressed liquefied nitrogen which gives their cooling effect to their own evaporation.

Immersion Freezing with Low Freezing Point Liquids

The low freezing point liquids that have been used for the non-packaged food includes solutions of sugars, sodium chloride and glycerol. These must be used at sufficient concentration to remain at -18 °C or lower to be effective. In the case of NaCl brine, this requires a concentration of 21%.

NaCl brine cannot be used with unpackaged food that should not become salty, especially fish and fish products. The disadvantage of NaCl brine is that the possible accumulation of the salt brine on the surface of the product. Sugar solutions have been used to freeze fruits at -18 °C and glycerol – water mixtures have been used to freeze fruits.

Immersion Freezing with Cryogenic Liquids

Cryogenic liquids are liquefied gases of low boiling point such as liquid nitrogen and liquid carbon dioxide with boiling point -196 °C and – 79 °C respectively. Today liquid nitrogen is the most commonly used cryogenic liquid in immersion freezing of food.

Advantages of Liquid Nitrogen

- It undergoes slow boiling at -196 °C, which provide a great driving force for heat transfer.

- Liquid nitrogen contacts all portion of irregularly shaped food, thus minimizing resistance to heat transfer.

- Since the cold temperature results from evaporation of liquid nitrogen, there is no need for a primary refrigerant to cool this medium.

- Liquid nitrogen is nontoxic and inert to food constituent. By displacing air from the food it can minimize oxidative changes during freezing and through packaged storage.

- The speed of liquid nitrogen freezing produces frozen food with a quality unattainable by non-cryogenic freezing methods. Some products such as mushroom cannot be frozen by other methods without excessive tissue damage. But such kinds of products can be effectively frozen by means of liquid nitrogen.

Advantages of Direct Immersion Freezing Methods

There is intimate contact between the food or package and refrigerant, so resistance to heat transfer is minimized. This is important with irregularly shaped food pieces to be frozen very rapidly such as loose shrimp, mushrooms and other food.

Although loose food pieces can be frozen individually by immersion freezing and air blast freezing, immersion freezing minimizing their contact with air during freezing, which can be desirable for food sensitive to oxidation. The speed of immersion freezing with cryogenic liquids produces quality unattainable by any other freezing method.

INDUSTRIAL FREEZE DRYING

Freeze-drying is a special form of drying that removes all moisture and tends to have less effect on a food's taste than normal dehydration process. In freeze-drying; food is frozen and placed in a strong vacuum. The water in the food sublimates, it turns straight from ice into vapor.

Freeze-drying is most commonly used to make instant coffee, but also works extremely well on fruits such as apples. Freeze-drying (also known as lyophilization or cryodesiccation) is a dehydration process typically used to preserve a perishable material or make the material more convenient for transport.

The alternative name for freeze drying favored by pharmaceutical industry is "Lyophilization". Freeze-drying works by freezing the material and then reducing the surrounding pressure and adding enough heat to allow the frozen water in the material to sublime directly from the solid phase to gas.

Freeze-drying Process

There are the following three stages in the complete drying process in industrial processes:

Freezing

The freezing process consists of freezing the material. In a lab, this is done by placing the material in a freeze-drying flask and rotating the flask in a bath, called a shell freezer, which is cooled by mechanical refrigeration, dry ice and methanol, or liquid nitrogen. On a larger-scale, freezing is usually done using a freeze-drying machine.

In this step, it is important to cool the material below the lowest temperature at which the solid and liquid phases of the material can coexist. This ensures that sublimation rather than melting will occur in the following steps. Larger crystals are easier to freeze-dry. To produce larger crystals, the product should be frozen slowly or can be cycled up and down in temperature.

This cycling process is called annealing. But large ice crystals will break the cell walls of the individual cells present in the food. Generally the freezing temperatures are between -50 °C and -80 °C. The freezing phase is the most critical in the whole freeze-drying process, because the product can be spoiled if badly done. Large objects take a few months to freeze-dry.

Primary Drying

During the primary drying phase, the pressure is lowered (to the range of a few millibars) and enough heat is supplied to the material for the water to sublimate. The amount of heat necessary can be calculated using the sublimating molecules' latent heat of sublimation. In this initial drying phase, about 95% of the water in the material is sublimated.

This phase may be slow (can be several days in the industry) and if too much heat is added, the material's structure could be altered. In this phase, pressure is controlled through the application of partial vacuum. The vacuum speeds sublimation, making it useful as a deliberate drying process.

Moreover a cold condenser chamber and condenser plates provide a surface for the water vapor to re-solidify on. This condenser plays no role in keeping the material frozen and it prevents water vapor from reaching the vacuum pump, which could degrade the pump's performance. Condenser temperature is typically below -50 °C (-60°F).

Secondary Drying

The secondary drying phase aims to remove unfrozen water molecules, since the ice was removed in the primary drying phase. This part of the freeze-drying process is governed by the material's adsorption isotherms. In this phase, the temperature is raised higher than the primary drying phase and can be above 0 °C, to break any physico-chemical interactions that have formed between the water molecules and the frozen material.

Usually, the pressure is also lowered in this stage to encourage desorption. However, there are products that benefit from increased pressure as well. After the freeze-drying process is complete, the vacuum is usually broken with an inert gas, such as nitrogen, before the material is sealed. At the end of the operation, the final residual water content in the product is extremely low.

Applications of Industrial Freeze-drying

The most important application of industrial freezing drying is explained below:

Pharmaceutical Industry and Biotechnology

Pharmaceutical companies often use freeze-drying to increase the shelf life of products, such as vaccines and other injectable. By removing the water from the material and sealing the material in a vial, the material can be easily stored, shipped and later reconstituted to its original form for injection.

Food Industry

Freeze-drying is used to preserve food and make it very light weight. The process has been popularized in the forms of freeze-dried ice cream; an example of astronaut food. The coffee is often dried by vaporization in a hot air flow, or by projection on hot metallic plates. Freeze-dried fruit is used in some breakfast cereal. Culinary herbs are also freeze-dried, although air-dried herbs are far more common and less expensive. However, the freeze-drying process is used more commonly in the pharmaceutical industry.

Technological Industry

In chemical synthesis, products are often lyophilized to make them more stable, or easier to dissolve in water for subsequent use. In bio separations, freeze-drying can be used also as a late-stage purification procedure, because it can effectively remove solvents. Also it is capable of concentrating substances with low molecular weights that are too small to be removed by a filtration membrane.

Other Uses

Freeze-drying creates softer particles with a more homogeneous chemical composition than traditional hot spray-drying, but it is also more expensive. Freeze drying is also used for floral preservation and wedding bouquet preservation has become very popular with brides who want to preserve their wedding day flowers.

Advantages of Industrial Freeze Drying

The most important advantages of freeze-drying are listed below:

- Minimum damage to the heat labile material.

- Creation of porous friable structures.

- Speed and completeness of rehydration.

- The ability to sterile filter liquids just before dispensing.

- The substance may be stored at room temperature without refrigeration and be protected against spoilage for many years. It would greatly reduce water content inhibits the action of microorganisms and enzymes that would normally spoil or degrade the substance.

- Freeze-drying does not usually cause shrinkage or toughening of the material being dried.

- The flavors, smells and nutritional content generally remain unchanged, making the process popular for preserving food.

- Freeze drying is useful when the product meets one or more of the following criteria:

  - It is unstable.

  - It is heat stable.

  - Quick and complete rehydration is required.

  - The product of high value.

  - Weight must be minimized.

  - Frozen or chilled storage is not appropriate.

Disadvantages of Freeze Drying

The principle disadvantages are:

- High capital cost of equipment.

- High energy cost.

- Lengthy process time (typically 4-10 hrs. per drying cycle).

- Possible damage to products due to change in pH and tonicity.

DEFROSTING FOODS

Preparing dinner shouldn't be a high-risk activity. Unfortunately, improper handling of frozen foods can pose a potential health risk to you.

When thawing frozen foods, make sure that the internal temperature of the food never gets into the "danger zone" of 40 °F and 140 °F. If it does, bacteria that cause food-borne illness can multiply extremely rapidly. Food should never be thawed at room temperature because the outer surface could easily warm to above 40° F while the center remains frozen.

There are three ways to safely defrost foods:

Refrigerator Method

This method involves placing a wrapped food item in a pan on the bottom shelf of the refrigerator away from other foods. Placing the food in a pan is important to prevent potential cross contamination from drippings as the product thaws.

Thawing this way can take several hours, so advance planning is necessary. For example, a large turkey can take three to four days to completely thaw in the refrigerator. Plan on a thawing time of four to five hours per pound for most foods when using this method.

Cold Water Method

A second way to safely thaw foods is to place a securely wrapped package of frozen food in a pan of cold water, changing the water approximately every 30 minutes. Continue the process until the food has thawed, which takes about 30 minutes per pound. Use this method only if you plan to cook the food immediately after thawing.

Microwave Method

Frozen foods can also be thawed in a microwave if you plan to cook the food immediately. Check your owner's manual for the minutes per pound and power level to use for thawing various types of foods. Be sure to rotate the food regularly to ensure even thawing. Any uncooked frozen foods thawed in a microwave must be cooked immediately.

Cooking from a Frozen State

If there isn't time to defrost, most foods can be cooked from the frozen state. If you cook food that is fully or partially frozen at the start, be sure to extend your cooking time to compensate and use a certified food thermometer to verify that the food has been cooked to a proper internal temperature throughout. Frozen foods can take up to 50 percent longer to cook than the thawed version of the same food.

FROZEN FOOD STORAGE

Golden Rules of Freezing Meals

To prevent those unwelcome discoveries in the back of your freezer, it helps to know how to freeze foods wisely in the first place. Keep in mind that when you freeze foods, you want to accomplish five things:

- Prevent freezer burn.

- Prevent moisture loss.

- Prevent the transfer of smells to and from other foods.

- Use what freezer space you have wisely.

- Prevent food poisoning as your food cools.

The key to accomplishing these goals lies in the proper wrapping and storing of your meals. Here are the golden rules for doing so:

- Leave as little air as possible in the freezer containers by removing as much air as possible from freezer bags before sealing them and by using freezer safe containers that closely fit the amount of food being frozen.

- Wrap meats and baked goods tightly with foil before you place them in freezer bags. Keep in mind that freezing meat in the packaging from the store (wrapped in plastic on Styrofoam trays) isn't ideal and won't hold up well to the freezer temperatures. You are usually okay if you use them within a month, however.

- To make sure your food freezes as quickly as possible to discourage bacteria growth, use small containers - with a capacity no bigger than 4 quarts. Ideally, the food should be less than 3 inches thick within the container.

- Cool your hot foods quickly before freezing them by placing the pan of hot food in a large container filled with ice or ice water, stirring often to keep the cold circulating. If you're cooling a lot of hot food, like a large saucepan of stew or chili, portion it into smaller, shallow containers.

- Label and date freezer bags or containers, even if you think you'll be using the contents within a week or two.

- Place the food items in the coldest part of your freezer, if you can, until they're completely frozen.

- Thawing food at room temperature only works with muffins, breads, and other baked goods. For everything else, thaw in the refrigerator or use the "thaw" setting on your microwave.

- Try to use your frozen foods within two to three months.

- When freezing dishes containing dairy foods, keep in mind that while milk can be frozen, it might separate a little when thawed. Hard and semi-hard cheeses can be frozen in 8- and 16-ounce blocks that have been wrapped in plastic, then put in freezer bags. While the cheese will still have its characteristic flavor when thawed, it could be a bit crumbly and tends to work best when added to cooked dishes. The cheeses that fare the worst with freezing are cream cheese and cottage cheese. Blue cheeses are most likely to become crumbly.

References

- Freezing-food-preservation: britannica.com, Retrieved 25 February, 2019

- Food-freezing-system-direct-and-indirect-food-technology-biotechnology, food-technology, food-biotechnology: biotechnologynotes.com, Retrieved 16 January, 2019

- Industrial-freeze-drying-process-applications-advantages-disadvantages-food-technology, food-technology, food-biotechnology: biotechnologynotes.com, Retrieved 29 March, 2019

- Defrosting, cooking, cooking-cleaning-food-safety, consumer-resources: nsf.org, Retrieved 30 April, 2019

- Frozen-food-storage-keeping-it-safe-and-tasty, features, food-recipes: webmd.com, Retrieved 29 June, 2019

Food Safety and Packaging

Food safety is concerned with handling, preparation and storage of food in such a way that it prevents food contamination. Food packaging refers to the enclosure of food to protect it from damage, spoilage and pest attacks. This chapter closely examines the key concepts of food safety and packaging as well as the materials which are commonly used to pack food.

FOOD SAFETY

Food Safety refers to handling, preparing and storing food in a way to best reduce the risk of individuals becoming sick from foodborne illnesses.

Food safety is a global concern that covers a variety of different areas of everyday life.

The principles of food safety aim to prevent food from becoming contaminated and causing food poisoning. This is achieved through a variety of different avenues, some of which are:

- Properly cleaning and sanitizing all surfaces, equipment and utensils.
- Maintaining a high level of personal hygiene, especially hand-washing.
- Storing, chilling and heating food correctly with regards to temperature, environment and equipment.
- Implementing effective pest control.
- Comprehending food allergies, food poisoning and food intolerance.

Regardless of why you are handling food, whether as part of your job or cooking at home, it is essential to always apply the proper food safety principles. Any number of potential food hazards exist in a food handling environment, many of which carry with them serious consequences.

Issues

Food safety issues and regulations concern:

- Agriculture and animal husbandry practices.

- Food manufacturing practices.

- Food additives.

- Novel foods.

- Genetically modified foods.

- Food label.

- Food Contamination.

Food Contamination

Food contamination happens when food are corrupted with another substance. It can happen. In the process of production, transportation, packaging, storage, sales and cooking process. The contamination can be physical, chemical and biological.

Physical Contamination

Physical contaminants (or 'foreign bodies') are objects such as hair, plant stalks or pieces of plastic and metal. When the foreign object comes into the food, it is a physical contaminant. If the foreign objects are bacteria, the case will a physical and biological contamination.

Common sources to create physical contaminations are: hair, glass or metal, pests, jewelry, dirt and fingernails.

Chemical Contamination

Chemical contamination happens when food is contaminated with a natural or artificial chemical substance. Common sources of chemical contamination can include: pesticides, herbicides, veterinary drugs, contamination from environmental sources (water, air or soil pollution), cross-contamination during food processing, migration from food packaging materials, presence of natural toxins or use of unapproved food additives and adulterants.

Chemical contaminations usually share the following characteristics:

- They are not intentionally added.

- Contamination can happen at one or more stages in food production.

- Illness is likely to result if consumers ingest enough of them.

Biological Contamination

Biological contamination refers to food that has been contaminated by substances produced by living creatures, such as humans, rodents, pests or microorganisms. This includes bacterial contamination, viral contamination, or parasite contamination that is transferred through saliva, pest droppings, blood or faecal matter. Bacterial contamination is the most common cause of food poisoning worldwide. If an environment is high in starch or protein, water, oxygen, has a neutral PH level, and maintains a temperature between 5 °C and 60 °C (danger zone) for even a brief period of time (~0-20 minutes), bacteria are likely to survive.

Example for Biological Contamination: Tainted Romaine Lettuce

Up to May, 2018, 26 states in the United States confirmed with an outbreak of the bacteria strain *E. coli* O157:H7. Several investigations show the contamination might have come from the Yuma, Ariz. growing region. This outbreak, which began April 10, is the largest US flare-up of E. coli in a decade. One person in California has died. At least 14 of the people affected developed kidney failure. The most common symptoms of E. coli include diarrhea, bloody diarrhea, abdominal pain, nausea and vomiting.

Safe Food Handling Procedures: From Market to Consumer

Proper storage, sanitary tools and work spaces, heating and cooling properly and to adequate temperatures, and avoiding contact with other uncooked foods can greatly reduce the chances of contamination. Tightly sealed water and air proof containers are good measures to limit the chances of both physical and biological contamination during storage. Using clean, sanitary surfaces and tools, free of debris, chemicals, standing liquids, and other food types (different than the kind currently being prepared, i.e. mixing vegetables/meats or beef/poultry) can help reduce the chance of all forms of contamination. However, even if all precautions have been taken and the food has been safely prepared and stored, bacteria can still form over time during storage. Food should be consumed within one to seven days while it has been stored in a cold environment, or one to twelve months if it was in a frozen environment (if it was frozen immediately after preparation). The length of time before a food becomes unsafe to eat depends on the type of food it is, the surrounding environment, and the method with which it is kept out of the danger zone.

- Always refrigerate perishable food within 2 hours—1 hour when the temperature is above 90 °F (32.2 °C).

- Check the temperature of your refrigerator and freezer with an appliance thermometer. The refrigerator should be at 40 °F (4.4 °C) or below and the freezer at 0 °F (-17.7 °C) or below.

For example, liquid foods like soup kept in a hot slow cooker (65 °C) may last only a few hours before contamination, but fresh meats like beef and lamb that are promptly frozen (-2 °C) can last up to a year. The geographical location can also be a factor if it is in close proximity to wildlife. Animals like rodents and insects can infiltrate a container or prep area if left unattended. Any food that has been stored while in an exposed environment should be carefully inspected before consuming, especially if it was at risk of being in contact with animals. Consider all forms of contamination when deciding if a food is safe or unsafe, as some forms or contamination will not leave any apparent signs. Bacteria may not be visible to the naked eye, debris (physical contamination) may be underneath the surface of a food, and chemicals may be clear or tasteless; the contaminated food may not change in smell, texture, appearance, or taste, and could still be contaminated. Any foods deemed contaminated should be disposed of immediately, and any surrounding food should be checked for additional contamination.

HAZARD ANALYSIS AND CRITICAL CONTROL POINTS

Hazard analysis and critical control points, or HACCP, is a systematic preventive approach to food safety from biological, chemical, and physical hazards in production processes that can cause the finished product to be unsafe and designs measures to reduce these risks to a safe level. In this manner, HACCP attempts to avoid hazards rather than attempting to inspect finished products for the effects of those hazards. The HACCP system can be used at all stages of a food chain, from food production and preparation processes including packaging, distribution, etc.

HACCP is believed to stem from a production process monitoring used during World War II because traditional "end of the pipe" testing on artillery shells' firing mechanisms could not be performed, and a large percentage of the artillery shells made at the time were either duds or misfiring. HACCP itself was conceived in the 1960s when the US National Aeronautics and Space Administration (NASA) asked Pillsbury to design and manufacture the first foods for space flights. Since then, HACCP has been recognized internationally as a logical tool for adapting traditional inspection methods to a modern, science-based, food safety system. Based on risk-assessment, HACCP plans allow both industry and government to allocate their resources efficiently in establishing and auditing safe food production practices. In 1994, the organization International HACCP Alliance was established, initially to assist the US meat and poultry industries with implementing HACCP, and now its membership has been spread over other professional and industrial areas.

Hence, HACCP has been increasingly applied to industries other than food, such as cosmetics and pharmaceuticals. This method, which in effect seeks to plan out unsafe practices based on science, differs from traditional "produce and sort" quality control methods that do nothing to prevent hazards from occurring and must identify them at the end of the process. HACCP is focused only on the health safety issues of a product and not the quality of the product, yet HACCP principles are the basis of most food quality and safety assurance systems. In the United States, HACCP compliance is regulated by 21 CFR part 120 and 123. Similarly, FAO and WHO published a guideline for all governments to handle the issue in small and less developed food businesses.

Principles

- Conduct a hazard analysis:

  Plan to determine the food safety hazards and identify the preventive measures the plan can apply to control these hazards. A food safety hazard is any biological, chemical, or physical property that may cause a food to be unsafe for human consumption.

- Identify critical control points:

  A critical control point (CCP) is a point, step, or procedure in a food manufacturing process at which control can be applied and, as a result, a food safety hazard can be prevented, eliminated, or reduced to an acceptable level.

- Establish critical limits for each critical control point:

  A critical limit is the maximum or minimum value to which a physical, biological, or chemical hazard must be controlled at a critical control point to prevent, eliminate, or reduce that hazard to an acceptable level.

- Establish critical control point monitoring requirements:

  Monitoring activities are necessary to ensure that the process is under control at each critical control point. In the United States, the FSIS requires that each monitoring procedure and its frequency be listed in the HACCP plan.

- Establish corrective actions:

  These are actions to be taken when monitoring indicates a deviation from an established critical limit. The final rule requires a plant's HACCP plan to identify the corrective actions to be taken if a critical limit is not met. Corrective actions are intended to ensure that no product is injurious to health or otherwise adulterated as a result if the deviation enters commerce.

- Establish procedures for ensuring the HACCP system is working as intended:

Validation ensures that the plants do what they were designed to do; that is, they are successful in ensuring the production of a safe product. Plants will be required to validate their own HACCP plans. FSIS will not approve HACCP plans in advance, but will review them for conformance with the final rule.

Verification ensures the HACCP plan is adequate, that is, working as intended. Verification procedures may include such activities as review of HACCP plans, CCP records, critical limits and microbial sampling and analysis. FSIS is requiring that the HACCP plan include verification tasks to be performed by plant personnel. Verification tasks would also be performed by FSIS inspectors. Both FSIS and industry will undertake microbial testing as one of several verification activities.

Verification also includes 'validation' – the process of finding evidence for the accuracy of the HACCP system (e.g. scientific evidence for critical limitations).

- Establish record keeping procedures:

The HACCP regulation requires that all plants maintain certain documents, including its hazard analysis and written HACCP plan, and records documenting the monitoring of critical control points, critical limits, verification activities, and the handling of processing deviations. Implementation involves monitoring, verifying, and validating of the daily work that is compliant with regulatory requirements in all stages all the time. The differences among those three types of work are given by Saskatchewan Agriculture and Food.

Application

- Fish and fishery products.
- Fresh-cut produce.
- Juice and nectar products.
- Food outlets.
- Meat and poultry products.
- School food and services.

FOOD PACKAGING

Automated palletizer of bread with industrial robots at a bakery.

Food packaging is the enclosing of food to protect it from damage, contamination, spoilage, pest attacks, and tampering, during transport, storage, and retail sale. The package is often labeled with information such as amount of the contents, ingredients, nutritional content, cooking instructions (if relevant), and shelf life. The package needs to be designed and selected in such a manner that there are no adverse interactions between it and the food. Packaging types include bags, bottles, cans, cartons, and trays.

Functions of Food Packaging

Food packaging serves many important functions. They may be broken down as follows.

- Containment: For items that are granulated, paper-based packages are the best, with a sealing system to prevent infiltration of moisture into the product. Other products are packaged using metal cans, plastic bags and bottles, and glass containers. Another factor in containment is packaging durability—in other words, the packaged food has to survive transport from the food processing facility to the supermarket to the home for the consumer.

- Protection: The packaging must protect the food from, (a) biological agents such as rats, insects, and microbes; (b) mechanical damage such as product abrasion, compressive forces, and vibration; and (c) from chemical degradation such as oxidation, moisture transfer, and ultraviolet light.

- Communication: Packaged food must be identified for consumer use, mainly with label text and graphics. It can also be done by using special shapes for the food package, such as the Coca-Cola bottle or the can of Spam. Other well-known food package shapes include potato chip bags and milk bottles. These packages also detail nutritional information, and whether they are packaged according to kosher or halal specifications. The label may also indicate whether it is safe to put the packaged food (such as a TV dinner) through a microwave process.

- Environmental issues: To protect the environment, we must be willing to reuse or recycle the packaging or reduce the size of the packaging.

- Package safety: Before using a particular type of package for food, researchers must ensure that it is safe to use that packaging for the food being considered, and that there are no adverse interactions between the package and the food. This includes any metal contamination issues from a can to the food product or any plastic contamination from a bottle to the food product.

- Product access: The packaging must be such that the product is readily accessible when the consumer is ready to use it. For example, pour spouts on milk cartons can make it easy to dispense the milk.

Food Packaging Types

The materials can be fashioned into different types of food packages and containers. Examples are given below.

Packaging type	Type of container	Examples of foods packaged
Aseptic packages	Primary	Liquid whole eggs
Plastic trays	Primary	Portion of fish
Bags	Primary	Potato chips
Bottles	Primary	Bottle of a soft drink
Boxes	Secondary	Box of soft drink bottles
Cans	Primary	Can of tomato soup
Cartons	Primary	Carton of eggs
Flexible packaging	Primary	Bagged salad
Pallets	Tertiary	A series of boxes on a single pallet, to transport packaged food from the manufacturing plant to a distribution center.
Wrappers	Tertiary	Used to wrap the boxes on the pallet for transport.

Primary packaging is the main packaging that holds the food that is being processed.

Secondary packaging combines the primary packages into a single box. Tertiary packaging combines all of the secondary packages into one pallet.

Special Techniques

- Vacuum packaging or inert atmosphere packaging: Oxygen in the air tends to reduce the shelf life of food by the process known as oxidation. To prevent this process, some foods are packaged at reduced pressure (partial vacuum) or using an inert gas (such as nitrogen) to replace oxygen.

- Bags-in-Boxes: These are used for soft drink syrups, other liquid products, and meat products.

- Wine box: This is a type of box designed for storage of wine.

Packaging Machines

The design and use of packaging machinery needs to take into account the following factors: technical capabilities, labor requirements, worker safety, maintainability, serviceability, reliability, ability to integrate into the packaging line, capital cost, floorspace, flexibility of use, energy usage, quality of outgoing packages, qualifications (for food, phamaceuticals, and so forth), throughput, efficiency, productivity, and ergonomics.

Packaging machines may be of the following general types:

- Blister, skin and vacuum packaging machines.
- Capping, over-capping, lidding, closing, seaming and sealing machines.
- Cartoning machines.
- Case and tray forming, packing, unpacking, closing and sealing machines.
- Check weighing machines.
- Cleaning, sterilizing, cooling and drying machines.
- Conveying, accumulating, and related machines.
- Feeding, orienting, placing, and related machines.
- Filling machines: Handling liquid and powdered products.
- Package filling and closing machines.
- Form, fill and seal machines.
- Inspecting, detecting and checkweighing machines.
- Palletizing, depalletizing, pallet unitizing and related machines.
- Product identification: For labeling, marking, and so forth.

- Wrapping machines.

- Converting machines.

- Other specialty machinery.

ACTIVE FOOD PACKAGING

The terms active packaging, intelligent packaging, and smart packaging refer to packaging systems used with foods, pharmaceuticals, and several other types of products. They help extend shelf life, monitor freshness, display information on quality, improve safety, and improve convenience.

The terms are closely related. *Active packaging* usually means having active functions beyond the inert passive containment and protection of the product. *Intelligent* and *smart* packaging usually involve the ability to sense or measure an attribute of the product, the inner atmosphere of the package, or the shipping environment. This information can be communicated to users or can trigger active packaging functions. Programmable matter, smart materials, etc can be employed in packages.

Depending on the working definitions, some traditional types of packaging might be considered as "active" or "intelligent". More often, the terms are used with new technologically advanced systems: microelectronics, computer applications, nanotechnology, etc.

Moisture Control

For many years, desiccants have been used to control the water vapor in a closed package. A desiccant is a hygroscopic substance usually in a porous pouch or sachet which is placed inside a sealed package. They have been used to reduce corrosion of machinery and electronics and to extend the shelf life of moisture sensitive foods and drugs.

Corrosion

Corrosion inhibitors can be applied to items to help prevent rust and corrosion. Volatile corrosion inhibitors (VCI) or vapor phase corrosion inhibitors can be provided inside a package in a pouch or can be incorporated in a saturated overwrap of special paper. Many of these are organic salts that condense on the metal to resist corrosion. Some films also have VCI emitting capability.

Films are available with copper ions in the polymer structure, These neutralize the corrosive gas in a package and deter rust.

VCIs create a neutral environment in the packaging. It works on the principle of difference in vapour pressure and causes reaction with metals and non-metals, and with moisture to prevent corrosion. There are different forms of VCIs available, such as papers, plastics, HDPE papers, oils, foams, chips, aluminum barrier foils, bubble, and emitters that can prevent corrosion at many stages.

Metal Chelation

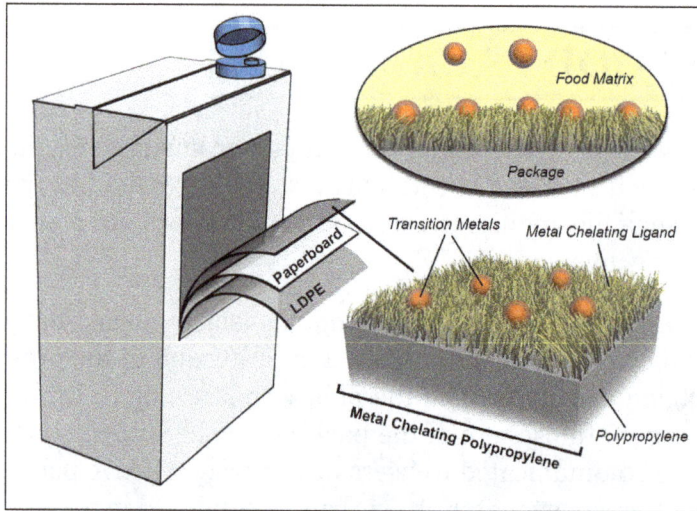

Synthesis of iminodiacetate functionalized polypropylene films and their efficacy as antioxidant active-packaging materials.

Trace transition metals in foods, especially iron, can induce oxidative degradation of many food components, especially lipids, and cause quality changes of the products. Metal-chelating active packaging materials are made by immobilizing metal-chelating active compounds onto traditional active packaging material. The surface immobilized metal-chelating compounds can scavenge the transition metals from the product and enhance the oxidative stability of the product. The metal-chelating active packaging technology is also antioxidant active packaging that will extend the shelf-life of consumer products by controlling the oxidation. The metal-chelating active packaging technology is known to be able to remove synthetic food preservatives (e.g. EDTA) from the food product. This technology can be used to address the increasing consumer demand for additive free and 'clean' label food products.

Oxygen Control

Oxygen scavengers or oxygen absorbers help remove oxygen from a closed package. Some are small packets or sachets containing powdered iron: as the iron rusts, oxygen is removed from the surrounding atmosphere. Newer systems are on cards or can be built into package films or molded structures. In addition, the physical characteristics of the packaging itself (oxygen transmission rate - OTR) can dictate how effective an oxygen absorber can be, and how long it will stay effective. Packaging with a low OTR will let less oxygen in the closed package through the polymer barrier itself.

Atmosphere

With some products, such as cheese, it has long been common to flush the package with nitrogen prior to sealing: the inert nitrogen is absorbed into the cheese, allowing a tight shrink film package. The nitrogen removes oxygen and interacts with the cheese to make the package functional.

More recently, other mixtures of gas have been used inside the package to extend the shelf life. The gas mixture depends on the specific product and its degradation mechanisms. Some package components have been developed that incorporate active chemistry to help maintain certain atmospheres in packages.

Oxygen scavengers, carbon dioxide generators, ethanol generators, etc. are available to help keep the atmosphere in a package at specified conditions.

Temperature Monitor

Some temperature indicators give a visual signal that a specified temperature has been exceeded. Others, Time temperature indicators, signal when a critical accumulation of temperature deviation over time has been exceeded. When the mechanism of the indicator is tuned to the mechanism of product degradation, these can provide valuable signals for consumers.

Digital temperature data loggers record the temperatures encountered throughout the shipment. This data can be used to predict product degradation and help determine if the product is suited for normal sale or if expedited sale is required. They also determine the time of the temperature excess: this can be used to direct corrective action.

Thermochromic inks are sometimes used to signal temperature excess or change. Some are reversible while others have a permanent change of color. These can be used alone or with other packaging functions such as barcodes.

The inks can also signal a desired temperature for consumers. For example, one type of beer can has ink that graphically shows when an ideal drinking temperature is achieved.

Controlling Package Temperatures

For critical vaccines, insulated shipping containers are passive packaging to help control the temperatures fluctuations seen even with a controlled cold chain. In addition, gel packs are often used to keep the temperature of the contents within specified acceptable temperature ranges.

Some newer packages have the ability to heat or cool the product for the consumer. These have segregated compartments where exothermic or endothermic reactions provide the desired effect. Self-heating cans are available for several products.

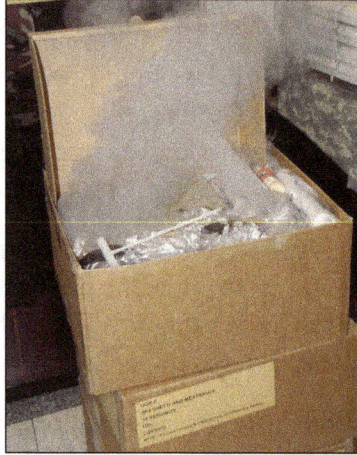

Self-heating field rations for up to 18 soldiers.

Dispensing

Some packages have closures or other dispensing systems that change the contents from a liquid to an aerosol. These are used for products ranging from precision inhalers for medications to spray bottles of household cleaners.

Some dispensing packages for two-part epoxy adhesives do more than passively contain the two components. When dispensed, some packages meter and mix the two components so the adhesive is fully functioning at the point of application.

The ability of a package to fully empty or dispense a viscous liquid is somewhat dependent on the surface energy of the inner walls of the container. The use of superhydrophobic surfaces is useful but can be further improved by using new lubricant-impregnated surfaces.

RFID

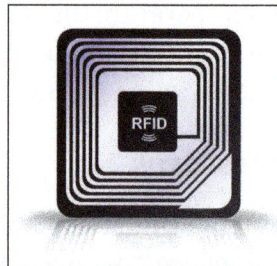

Radio-frequency identification chip.

Radio-frequency identification chips are becoming more common with the introduction of smart labels that are used to track and trace packages and unit loads throughout distribution. Newer developments include recording the temperature history of shipments and other intelligent packaging functions. RFID can be integrated into labels: Smart labels.

Security

A variety of security printing methods, security holograms, and specialized labels are available to help confirm that the product in the package is not counterfeit. RFID chips are being used in this application also.

Electronic surveillance (on the product or on the package) is used to help counter shoplifting.

Microwave Packaging

Metallised films are used as a susceptor for cooking in microwave ovens. These increase the heating capacity and help make foods crisp and brown. Plastic microwavable containers are also used for microwave cooking.

Shock and Vibration

Shock detectors have been available for many years. These are attached to the package or to the product in the package to determine if an excessive shock has been encountered. The mechanisms of these *shock overload* devices have been spring-mass systems, magnets, drops of red dye, and several others.

Recently, digital shock and vibration data loggers have been available to more accurately record the shocks and vibrations of shipment. These are used to monitor critical shipments to determine if extra inspection and calibration is required. They are also used to monitor the types of shocks and vibrations encountered in transit for use in package testing in a laboratory.

Other Developments

Chlorine dioxide pouches placed inside fruit-packing boxes kill pathogens but don't damage fruit.

Some engineered packaging films contain enzymes, anti-bacterial agents, scavengers, and other active components to help control food degradation and extend shelf life. Edible films have been developed to allow consumers to eat the package along with the product.

Special packaging has been developed for shipping organs which keeps them alive during extended shipments. The organs are alive and fresh for transplanting.

Several packages used by Canadian cannabis corporations use active packaging to monitor THC levels throughout the production process. This is being implemented in order to ensure consistency between products to improve supply chain management as well as offer consumers improved value of purchase.

Regulations

Active packaging is often designed to interact with the contents of the package. Thus extra care is often needed for active or smart packagings that are food contact materials.

Food packagers take extra care with some types of active packaging. For example, when the oxygen atmosphere in a package is reduced for extending shelf life, controls for anaerobic bacteria need to be considered. Also when a controlled atmosphere reduces the appearance of food degradation, consumers need to retain a means of determining whether actual degradation is present.

TYPES OF PACKAGING MATERIAL USED IN FOOD

Packaging and containers have become an essential element in food purchases. The food is packaged and packed with the aim of being transported and stored. That is, not only is it a container, but "the container must protect what it sells and sell what it protects".

From the business perspective, the appearance of packaging is particularly important because it identifies the product in the distribution chain and differentiates it when it reaches the consumer.

Next, the most used materials in the food industry are detailed: plastics, glass, metals and wood and its derivatives.

Plastics

Plastics are organic polymeric materials that can be molded into the desired shape. The lightness and versatility of these have been confirmed over decades in the processing and packaging of food. Plastic containers and packaging protect against the contamination of food and offer adequate mechanical strength.

Due to a lower cost and lower energy consumption during manufacturing, plastics have replaced traditional packaging materials. In addition, they are able to preserve and protect the food for longer, minimizing the use of preservatives.

In relation to the consumer, they are easy to handle and open, and offer an effective surface for printing labels or brands. However, although plastics are recyclable materials, they are pollutants.

In the plastic manufacturing process, there are many varieties of plastic resins, with the most used being:

- Polyvinyl chloride (PVC): It is very resistant to humidity, fats and gases.

- Polyethylene and its varieties (PET, HDPE, LDPE): The development of the PET range has revolutionized the packaging industry, allowing plastic to compete directly with glass bottles.

- Polystyrene (PS): It is the plastic of choice for thermoforming due to its strength, malleability and low cost.

Cellulose is a biodegradable substance obtained from the cell wall of many vegetables and fungi. It was the first transparent film that was used in packaging and is currently used for confectionery and pastry products, in situations where vapors need to "breathe" to avoid deforming the product.

Polyamides are a type of polymer that can be found in nature such as wool, or synthetically, like nylon. They are used for boiled products in bags, frozen foods, fish, meat, vegetables and processed meat and cheese.

Metals: Steel, Tin and Aluminum

The main use of these metals is the preservation of canned foods and beverages. The most commonly used are tin-coated steel and aluminum cans. It is an opaque material that provides an advantage for food that is sensitive to light.

Tin cans are made of steel sheet coated with tin as a measure of protection against corrosion of steel, especially when they contain products with low pH.

Aluminum is increasingly used for canning due to its lightness, low cost and capacity to be recycled. It can be found in packaging, bottle closures and wraps and laminates. It has the same barrier properties as steel but with the advantage of being resistant to corrosion.

Aluminum foil is formed by layers of laminated aluminum. It is a highly flexible product that allows to preserve or protect food in the domestic environment. However, it is difficult to use in modern fast packaging equipment due to wrinkles, rips and marks.

Thin-walled aluminum cans are suitable for carbonated beverages, while wide-walled cans are suitable for steam sterilization. Optionally, internal lacquers can be used to avoid interaction with the product and externally to protect the ink from the labeling.

Glass

Glass is an inert material that is impermeable to gases and vapors. It is an excellent and completely neutral oxygen barrier when in contact with food. However, it is a fragile, heavy material that requires a lot of energy to be manufactured.

Glass uses one of the most abundant raw materials on the planet, silica, but it is not renewable. Despite this, it is a recyclable product, since it can be used as a container repeatedly.

More than 75 billion glass containers are used per year in the food industry, being its main use for wines, juices, baby food and soft drinks.

Glass containers can be bottles (the most used), jars, glasses, ampoules, jars, etc. However, this material is not used for frozen products due to the risk of breakage.

Wood, Cardboard and Papers

Products derived from wood are widely used in the packaging of food in the form of paper and cardboard. Paper is a very cheap, lightweight product with excellent printing capacity. Although it is very sensitive to moisture, it can be corrected with a combination of paper and other materials such as plastic or paraffin.

Cardboard is a material made up of several superimposed layers of paper, making it thicker, harder and more resistant than paper. Its main use is for packaging and containers in the form of boxes.

In recent years, paper and cardboard manufacturers are paying special attention to issues related to health and the environment by working with recycled products that increase the useful life of these raw materials.

Advances in Packaging Techniques

The evolution of packaging techniques, together with those of food preservation, transform the processes of manufacturing, distribution, purchase and preparation of food, for both businesses and consumers.

Technological advances, such as the incorporation of antioxidants in food packaging, increase the shelf life of food. This system is based on the addition of particles to the packaging materials to prevent the oxidation of nutrients. In this case, the antioxidants can be incorporated during the manufacturing process or impregnate the walls of the container before coming into direct contact with the food.

In short, it is important to choose the appropriate packing and container material for each food to be conserved, taking into account the transport circumstances and storage conditions to which it is going to be subjected.

SMART OR INTELLIGENT PACKAGING

Smart or intelligent packaging is another popular method of food packaging. Intelligent packaging is designed to sense the environment and convey information to the user. It helps to monitor and communicate information about food quality. Such packaging has gained prominence for tracking and tracing perishable food commodities. There are various features in intelligent food packaging, like as follows:

- Radio frequency identification.

- Ripeness indicators.

- Time-temperature indicators.

- Biosensors.

THE GROWING IMPORTANCE OF FOOD PACKAGING

Food packaging is very important not only as a protection of food but also as a means of marketing the food product. Based on the latest advanced technology, food manufacturers, retailers, packaging products suppliers are working more closely than ever before to produce packaging products designed for modern lifestyles. This has led to the emergence of varied hi-fi packaging designs and packaging products, especially for food packaging, in the market. There has been explosion of ready to eat meals in the market. All food packaging highlights freshness, health and traceability of the products. Packaging has become an important part of the value chain analysis, regarding food safety.

The importance of food packaging is further strengthened by the final choice of the consumers because it directly involves appeal, convenience, information and branding. The food industry and the food packaging industry is reinventing the food service channel.

Proper Food Packaging Functions

Food packaging serves several purpose:

- Protection: The food kept in the package require protection from various things like vibration, shock, compression, temperature, etc.

- Barrier protection: Packaging is required to provide a barrier from oxygen, dust, water vapor, etc. Keeping the contents of the food fresh, clean, and safe for the required shelf life is a basic function.

- Containment: Small food items are typically grouped together in one package for efficiency. Granular materials and powders need containment.

- Label Information: All food products packages require labels on the nutrition facts, how to use, dispose, date of manufacturing and expiry of the package or product.

- Marketing: Proper food packaging with proper information plays a vital role in encouraging potential customers to buy the product and this case the food packaging design plays an equally important role. The more attractive the design is, the more is its chances of buying.

- Security: Proper packaging can play an essential role in reducing the security risks of shipment. With tamper evident features, there is no chances of the food packages being destroyed or contaminated during shipment.

- Convenience: Food packages have certain features which add convenience in handling, stacking, distribution, sale, display, re-closing, opening, use, and reuse.

- Portion control: Lastly, food packaging provides information on the exact portion of contents available in that particular product.

MATERIALS FOR FOOD PACKAGING

Aluminium

More than 90% of all food products sold in Western Europe come packaged. The packaging not only protects the food from spoilage and contamination, but also often is chosen in order to improve the looks so that more people of the respective target group are drawn to the product.

AlFiPa supplies aluminium foils and laminates for numerous applications of food packaging. Whether it is foils, tapes or cuttings, you will find everything that you are looking for.

Benefits of Aluminium foil when Packaging Food

Aluminium packaging is light, flexible and easily recyclable. Furthermore, it is hygienic, non-toxic and helps in keeping the aroma of foods. It keeps the food fresh for a long time and provides protection from light, ultraviolet radiation, oils and grease, water vapour, oxygen and microorganisms.

Lasagne in an Aluminium container.

In addition, have a look at the following advantages:

- Aluminium foil reacts only to highly concentrated acids and basic substances and is otherwise strongly corrosive resistant.

- Aluminium foil is sterile and therefore hygienic.

- Aluminium foil is tasteless and odorless.

- Aluminium foil is extremely dimensionally stable even in soft state.

- Aluminium foil can be recycled several times without loss of quality.

- Aluminium foil does not absorb liquids.

Despite the contrary claims, aluminum foil in food packaging is generally harmless to health. Acid or salty foods should not come into direct contact with aluminium, but composite films with aluminum layer can be used.

The efficient storage of food and beverages in aluminum packaging reduces the need for cooling and makes it spoil slower. In addition, the low weight of the aluminum reduces required energy during transport.

Aluminum foils and laminates are ideally suited for the production of packaging for:

- Coffee and tea.

- Instant beverages.

- Jams and pies.

- Snacks, biscuits and peanuts.

- Sausage and dried meat.

- Soup concentrates.

- Baby food.

- Ready meals and preserves.

The efficient storage of food and beverages in aluminum packaging reduces the need for cooling and makes it spoil slower. In addition, the low weight of the aluminum reduces required energy during transport.

Aluminum foils and laminates are ideally suited for the production of packaging for:

- Coffee and tea.

- Instant beverages.

- Jams and pies.

- Snacks, biscuits and peanuts.

- Sausage and dried meat.

- Soup concentrates.

- Baby food.

- Ready meals and preserves.

Colorful Packaging with Aluminum Foil

Colorful packages often sell better. In many cases, costumers choose a particular product because they feel attracted by the design of the packaging.

Food packaging made of Aluminium.

As in most other industries, colored films are often used in the food industry too. Whether imprinted or painted – food is particularly effectively sold in colorful looks. With visually successful food packaging, manufacturers can give their products their own image and thus promote the recognition value of the target group.

Which Aluminum Foil is Suitable for which Product?

Depending on their thickness, aluminum foils and laminates are used for a wide range of food packaging.

Films of 10 to 12 µm are ideal for chocolates and candies, while 30-38 µm films are used in the dairy industry, for example for packaging desserts, pudding and yogurt. For marmalades and pies, however, aluminum foils of 50-70 µm are the right choice.

Due to cost reduction, the trend is currently going towards increasingly thinner aluminium foils. In the field of chocolate packaging, the thickness of the packaging alone decreased by about 30% in the last twenty years. Currently, the average thickness of chocolate foil is between 7-15 µm, less than the diameter of a human hair.

Bare Aluminum Foil is not Suitable for all Foodstuffs

The selection of a suitable packaging film helps to protect the product optimally and to extend the minimum shelf life.

However, in the case of aluminum foil it is only guaranteed if it does not come into contact with particularly acidic or salty foods. Under the influence of acid and salt, aluminium ions dissolve and pass into the food.

It is therefore not recommended to use bare aluminum foils as packaging for the following products:

- Apple sauce, stewed apples, apple purée.
- Rhubarb.
- Tomatoes.
- Sour cucumbers.
- Sauerkraut.
- Salted herring.
- Sliced citrus fruits.
- Marinades that contain vinegar or fruit acids.
- Lye pastry before baking.

However, laminates are well suited for packaging those products. Composite films can be used – also with aluminum. By protecting the aluminum with a plastic layer the food does not come into direct contact with the foil.

Stand-up pouch made an Aluminum composite foil.

Contact with other metals also leads to corrosion of aluminum soil. This effect is noticeable by a dark to black coloring of the film, which could even partly dissolve.

Plastics

Not-so-fantastic Plastic

Though the risk is low, there's growing evidence that food can be contaminated by harmful chemicals from some types of plastic. Many foods are packaged in these risky plastics – including fresh meat, gourmet cheese, and even some health foods and organic vegetables.

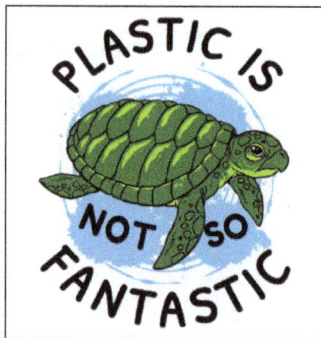

Problem with Plastic

Plastic as such isn't a problem. The polymer molecules from which it's made are far too big to move from the packaging material into the food. But plastic can also contain much smaller molecules that are free to migrate into the food it's in contact with. The plastic itself can slowly break down, releasing monomer, or other chemicals may be added to the plastic to give it the right mechanical properties. Two plastics of particular concern are:

- Polycarbonate – It is often used to make food storage containers and bottles, and the epoxy resin used to line cans. It can release bisphenol A (BPA), a chemical that many experts now believe can cause serious health problems.

- PVC – It is used to make bottles, cling wrap and the seals for screw-cap jars. On its own, PVC is hard and rigid (it's used to make drains, guttering and downpipes), so extra chemicals called plasticisers are added to make it soft and flexible – in much the same way water added to clay makes it soft. Plasticisers can make up as much as 40% of the plastic material. Phthalates and epoxidised soybean oil (ESBO) are often added as plasticisers to the PVC that's used for food packaging.

Risk Involved

BPA and some phthalates are endocrine disruptors, meaning they can mimic the body's natural hormones and thereby cause a raft of health problems. Infants and the very young are most vulnerable to exposure because of their lower body weight and because their growth and development are strongly influenced by hormones; the effects on health can be lifelong. These effects have been seen clearly and consistently in experiments with animals, and when people or wildlife have been accidentally exposed to high levels of endocrine disruptors.

While these compounds are undoubtedly hazardous at high levels of exposure, scientific opinion is divided over the risk from the much lower levels that we're exposed to every day in our food. There is, however, growing scientific evidence that even at these lower levels of exposure, phthalates and BPA may be causing problems such as infertility, obesity, breast cancer, prostate cancer, heart disease and diabetes.

BPA

BPA is rapidly eliminated from the body, but because of continuous exposure most of us have detectable levels of BPA in our body tissue. Typical levels, however, are well below the daily upper limit of safe exposure set by the US Food and Drug Administration and the European Food Safety Authority. But many independent scientists have expressed concern that this limit is based on experiments done in the 1980s, rather than on the hundreds of more recent animal and laboratory studies suggesting we could be at risk from much lower doses. Such low dose effects now have enough scientific credibility for the American Medical Association (AMA) to call on the US government to enact new federal policies to decrease the public's exposure to endocrine-disrupting chemicals. In particular, the AMA stated that, "Policy should be based on comprehensive data covering both low-level and high-level exposures."

Not surprisingly, the plastics industry strenuously refutes these findings and continues to insist that BPA is harmless at the low levels to which we're regularly exposed in our food. But while the evidence is far from conclusive, there's now far too much of it to be ignored. The underlying science is sound and the potential for such effects is real.

Phthalates

Phthalates are now used in so many products they are almost impossible to avoid. A Swiss study found people who eat healthily and try to avoid chemical additives in their food are exposed to much the same levels of phthalates as those who eat junk food and don't worry about their diet at all. Experiments with animals have consistently shown that some phthalates can be endocrine disruptors but, as with BPA, the evidence for adverse health effects from low-level exposure to phthalates is more limited. Again, though, there's too much of it to be ignored.

Because of its low cost, DEHP is the phthalate most often used as a plasticiser for PVC. Experts now generally agree that low level exposure to DEHP can affect reproductive development, particularly in young boys, and a US study has found a link between exposure to phthalates and increased risk of diabetes and obesity in men.

ESBO

ESBO is one of the most frequently used additives to PVC when used for containers or packaging for food. It functions as a stabiliser as well as a plasticiser. Lid seals are formed at high temperatures, which causes the PVC in the seal to partially break down and release hydrogen chloride.

ESBO reacts with the hydrogen chloride and prevents further breakdown of the plastic, but in doing so it forms compounds called chlorohydrins. Chlorohydrins make up, at most, five per cent of the ESBO but they can be toxic. Chlorohydrins have been detected in foods closed in glass screw-cap jars.

Types of Plastics and their Uses

Uses of Polyethylene terephthalate (PET):

- Bottles used for water and softdrinks.

- Jars for products such as peanut butter.

- Lightweight and 'green' wine bottles.

Uses of High density polyethylene (HDPE):

- Bottles used for milk and cream.

- Yoghurt cups.

- Bags that line breakfast cereal packets.

Uses of Polyvinyl chloride (PVC):

- Shrink and cling wrap.

- Clear plastic containers for fresh fruit or takeaway sandwiches.

- Some soft drink bottles.

- The gaskets that form a seal on screw-cap glass jars.

Uses of Low density polyethylene (LDPE):

- Take-away containers.

- Waterproof coating on milk cartons.

- Bags used for bread and frozen foods.

- Cling wrap.

Uses of Polypropylene (PP):

- Bottle caps.

- Yoghurt and margarine containers.

- Food storage boxes.

Uses of Polystyrene (PS):

- Plastic cutlery.

- Drinking cups and yoghurt cups.

- Cups for hot coffee (polystyrene foam).

- Lightweight trays used by supermarket to package and sometimes vegetables (polystyrene foam).

What are the Regulators Doing?

The plastics industry has been fighting off tighter regulation. It's a huge industry with vast resources (worldwide, it produces about 0.4 million tonnes per year of phthalates and more than two million tonnes of BPA) and independent scientists have complained about an aggressive disinformation campaign. Certainly, industry websites blatantly highlight studies that support their point of view and ignore those that don't.

In 2008 the Productivity Commission recommended that the Australian government establish a more systematic research program to identify and deal with the risks of chemicals in consumer articles, but to date there's been little action. Our regulators could do more to protect consumers; a lack of evidence of harm is not evidence of safety.

The use of plastics for wrapping or packaging foods is governed by the Food Standards Code, which sets a limit for the level permitted in food of highly toxic vinyl chloride monomer (10 parts per billion) yet no specific limits for BPA, DEHA or phthalates. These compounds come under a vague clause in this code that prohibits materials "likely to cause bodily harm, distress or discomfort". Food Standards Australia New Zealand (FSANZ), our food regulator, maintains that BPA and phthalates pose no significant health risks at the low levels found in food.

Consumers in Europe and North America are better protected.

Canada, the European Union and some states of the US have phased-out the use of BPA in some products. In the US at the federal level, the FDA is taking steps to reduce human exposure to BPA in the food supply. These steps include:

- Supporting the industry's actions to stop producing BPA-containing baby bottles and infant feeding cups for the US market.

- Facilitating the development of alternatives to BPA for the linings of infant formula cans.

- Supporting efforts to replace BPA or minimise BPA levels in other food can linings.

In Australia, nothing has been done other than the introduction of a purely voluntary phase-out by major retailers of polycarbonate plastic baby bottles containing BPA.

The European Union began to take action on phthalates in 1999. As a result six phthalates (including DEHP) have been banned in toys and other children's products at levels greater than 0.1%. The EU has also restricted the use of these phthalates in food contact applications. Since 2008 the US has banned DEHP and other phthalates at levels greater than 0.1% in toys and childcare articles.

In Australia, the National Industrial Chemicals Notification and Assessment Scheme (NICNAS) reported on DEHP in 2010 and recommended action. DEHP is now banned from toys and childcare articles, but only at levels exceeding one per cent – a limit 10 times higher than in the US and the EU.

ESBO

While there's no evidence that ESBO itself is harmful, an expert committee appointed by the European Union (EU) to review the evidence about ESBO concluded that "in the absence of adequate analytical and toxicological data on ESBO derivatives, no advice can yet be given on the significance for health of such derivatives in foods".

In other words, we can't be sure that our health is not being put at risk from the use of ESBO.

Plastic Products to Avoid

Most food and drink is packaged in containers made from plastics that seem to be harmless. Soft drinks and bottled water are usually in PET bottles, for example, while yoghurt and margarine containers are usually made from polypropylene. There's clearly no real need for food manufacturers to use packaging or wrapping made from potentially harmful plastics like polycarbonate or PVC, but there are still far too many products in the supermarkets where the food is in contact with them.

- You can often identify the type of plastic from its identification code – unfortunately, this code is voluntary and you won't find it on all plastic packaging. Look for the codes 1 (PET), 2 (HDPE), 4 (LDPE), 5 (PP) and 6 (PS). Whenever possible avoid the codes 3 (PVC) or 7 (a catch-all category that includes polycarbonate).

- Avoid fresh meat, fruit or vegetables wrapped in cling wrap. Most cling wrap sold for domestic use is now made from low density polyethylene, which seems to be safe, but supermarkets and many independent butchers and greengrocers are still wrapping meat and fresh vegetables in cling wrap made from PVC.

- Avoid reusable plastic bottles with the symbol 7 (or look for product labels that say "BPA-free"). Keep in mind that heating and washing polycarbonate bottles can increase the amount of BPA that leaches out.

- Consider cutting down on canned foods, as can linings can leach BPA directly into food.

- While some plastics such as polypropylene (often used for take-away containers) seem to be OK, as a general rule it's probably safer to avoid using any plastic containers when cooking or reheating food in a microwave oven. Use glass containers for high-fat foods, as toxic chemicals are more likely to migrate into fatty foods at high temperatures.

Bamboo Packaging

Bamboo is a common commodity that has been used by humans for at least 5,000 years. In China, it's a symbol of uprightness and in India a symbol of friendship. Bamboo's uses span countless industries, like construction, textiles, food production and even musical instruments. It is a fast-growing, sustainable source of material that helps businesses reduce environmental impact and promote a cleaner, greener planet.

Bamboo has also found its way into the cosmetics industry, as a source of luxurious and sustainable packaging for natural cosmetics.

Bamboo is a woody plant that is actually a member of the grass family. Bamboos are some of the fastest growing plants on the planet, with some being able to grow almost 4 centimeters per hour. This rapid growth characteristic, as well as it's ability to be used for construction and culinary purposes has led to bamboo being of significant cultural and economic importance to many Asian countries.

Use of Bamboo Packaging

Using bamboo packaging for cosmetics has numerous benefits for producers, consumers, as well as the planet! You can feel confident that your bamboo packaged products are the best option for the planet because:

- Strength and Durability – Bamboo is a durable material that can withstand great amounts of stress, having mechanical properties 2 to 3 times better than traditional timber. Your sustainable bamboo packaging can and should be reused.

- Quick Growing – Bamboo grows much faster than real wood, making it much more renewable and requiring less land and resources to produce. Among the world's fastest growing plants, bamboo can grow one inch (2.5 centimeters) every 40 minutes.

- Locally Sourced – Bamboo grows in many different parts of the world, and as a result, packaging produced with bamboo is often locally sourced. This is much more eco-friendly, limiting transportation costs and carbon footprints.

- Earth Friendly – Bamboo is a hardy, easy-to-grow grass that helps promote healthy soil and doesn't require replanting after harvesting. Bamboo packaging is biodegradable and can be composted.

Glass Packaging

There's just something so much better about using glass. Whether it's glass bottles, jars, or square containers, there are so many advantages to this material.

Hidden Advantage of Glass Packaging

We all want our products to be the best that they can be. But it's just as important to make sure the outside of the product matches the quality on the inside. The right packaging can make a significant difference for all around experience on all ends of the spectrum.

Protect your brand and keep your consumers happy by choosing glass packaging for all your dairy products. Milk, yogurt, cottage cheese anything goes in glass.

Delicious Taste comes from Glass Bottles

Glass bottles preserve the flavor of your dairy product much longer than other types of packaging. This is because the material is less likely to allow contents to mix with air or other possible chemicals.

Moisture is also less likely to get in the container, keeping your food safe from mold or going stale faster than you'd like it to.

While plastic erodes over time, glass has to physically break to become contaminated. This protects your product and keeps your consumers coming back for more delicious taste.

80 percent of consumers in a survey said that they believe that glass containers maintain the quality of the product better than other packaging. Since consumers are aware that glass is better for taste, they're more likely to keep buying your product in glass.

Glass Packaging is Healthier and Safer for Consumers and their Families

Glass is always a safe choice to use for packaging products. Plastic can melt, and there are risks of chemicals leaking into your food. Certain plastics aren't even safe for food production.

Plastic containers are at risk for high levels of BPA, which is a compound that is used to make certain types of plastic. BPA can seep in through the food containers and contaminate the product within that container.

BPA has serious health consequences when ingested, especially for children and babies. It can affect hormone levels, brain function, and prostate function. With glass, you don't have to stress about chemicals and contaminants. Glass is made from natural materials including sand, limestone, and soda ash. It's actually the only packaging material that has been designated as fully safe by the FDA.

Glass is chemically inert because of its natural composition, which means it isn't made up of reactive chemicals. There is a high barrier between different chemicals and substances. This high barrier means that your product is less likely to become contaminated, and it will stay fresh for longer.

The world is full of health scares and terrifying food contamination stories. Don't let your brand become another news story. Keep your customers safe and healthy so they'll keep coming back for years to come.

Surface Value of Dairy Products in Glass Containers

One of the most obvious benefits that comes from putting your dairy products in glass containers is the visual appeal. It's much more sophisticated than plastic or paperboard and can really make a shelf look great.

Glass bottles are attractive to customers because they have a classic look. They never get old or boring. They stay simple and never go out of style.

Many people also choose to keep and repurpose their glass containers once they're done using the product. There's a reason they don't do this with plastic jugs- they don't look very nice.

You have the opportunity to create a beautiful, colorful label while keeping the bottle simple and traditional. It's not confusing and your customers will know exactly what they're getting.

Simplicity is key when people are buying food products because they want to be able to find what they're looking for on the package. The majority of today's consumers read the label of a product, and if they don't find what they're looking for they won't buy it.

Easier Consumer Experience with Glass

When a plastic product is used over time, the plastic becomes distorted. It gets bent inward on certain sides, or shifts its shape to an inconvenient form.

This can cause issues when your customers are trying to get the last little bit out, or trying to scoop the product out.

Glass always stays the same. It's solid and will not lose its shape regardless of how much you try to bend it or twist it. When it's time to get down to the last little bit, it's easy to get around because the container won't move.

Environmental Benefits from Glass Dairy Packaging

Let's not forget how beneficial glass packaging is for the environment. In today's world, environmental consciousness is a huge deal and a big selling factor for many consumers.

Glass is completely recyclable, and it can be used over and over again as many times as necessary. It is the only packaging product on the market that has an endless recycling life.

It also has one of the fastest turnover times when it's recycled. A glass bottle can go from the recycling bin, through the recycling process, and be made into a new bottle ready for use within 30 days.

Many companies will also collect those containers when you're done with the product and take them back to reuse again. For example, with home milk delivery in glass bottles, the delivery company will take back the empty bottles and replace them with new, full ones.

Consumers can also reuse the glass bottles or containers as many times as they wish. There are many crafts and decorations that people can do using glass bottles.

References

- Food-packaging: newworldencyclopedia.org, Retrieved 14 July, 2019

- CIEH. "Chartered Institute of Environmental Health (CIEH)". Archived from the original on 25 February 2009. Retrieved 25 March 2009

- What-is-food-safety: foodsafety.com.au, Retrieved 11 January, 2019

- "Prevention of foodborne disease: Five keys to safer food". World Health Organisation. Retrieved 10 December 2010

- Aluminum-foil-laminates-food-packaging, applications: alfipa.com, Retrieved 27 May, 2019

- Soroka, W (2008). Illustrated Glossary of Packaging Terms. Institute of Packaging Professionals. P. 3. ISBN 978-1-930268-27-2

- Packaging-material-food: btsa.com, Retrieved 21 July, 2019

- Active-intelligent-food-packaging: packagingconsultancy.com, Retrieved 18 April, 2019

Processing of Common Food

The conversion of one form of food into another form is known as food processing. Some of the common foods which undergo processing before consumption are cereals, meats and milk. The topics elaborated in this chapter will help in gaining a better perspective about processing of these foods.

MILK PROCESSING

Milk is a nutritive beverage obtained from various animals and consumed by humans. Most milk is obtained from dairy cows, although milk from goats, water buffalo, and reindeer is also used in various parts of the world. In the United States, and in many industrialized countries, raw cow's milk is processed before it is consumed. During processing the fat content of the milk is adjusted, various vitamins are added, and potentially harmful bacteria are killed. In addition to being consumed as a beverage, milk is also used to make butter, cream, yogurt, cheese, and a variety of other products.

The use of milk as a beverage probably began with the domestication of animals. Goats and sheep were domesticated in the area now known as Iran and Afghanistan in about 9000 B.C., and by about 7000 B.C. cattle were being herded in what is now Turkey and parts of Africa. The method for making cheese from milk was known to the ancient Romans, and the use of milk and milk products spread throughout Europe in the following centuries.

Cattle were first brought to the United States in the 1600s by some of the earliest colonists. Prior to the American Revolution most of the dairy products were consumed on the farm where they were produced. By about 1790, population centers such as Boston, New York, and Philadelphia had grown sufficiently to become an attractive market for larger-scale dairy operations. To meet the increased demand, farmers began importing breeds of cattle that were better suited for milk production. The first Holstein-Friesens were imported in 1795, the first Ayrshires in 1822, and the first Guernseys in 1830.

With the development of the dairy industry in the United States, a variety of machines for processing milk were also developed. In 1856, Gail Borden patented a method for making condensed milk by heating it in a partial vacuum. Not only did his method remove much of the water so the milk could be stored in a smaller volume, but it also protected the milk from germs in the air. Borden opened a condensed milk plant and cannery in Wassaic, New York, in 1861. During the Civil War, his condensed milk was used by Union troops and its popularity spread.

In 1863, Louis Pasteur of France developed a method of heating wine to kill the microorganisms that cause wine to turn into vinegar. Later, this method of killing harmful bacteria was adapted to a number of food products and became known as pasteurization. The first milk processing plant in the United States to install pasteurizing equipment was the Sheffield Farms Dairy in Bloomfield, New Jersey, which imported a German-made pasteurizer in 1891. Many dairy operators opposed pasteurization as an unnecessary expense, and it wasn't until 1908 that Chicago became the first major city to require pasteurized milk. New York and Philadelphia followed in 1914, and by 1917 most major cities had enacted laws requiring that all milk be pasteurized.

One of the first glass milk bottles was patented in 1884 by Dr. Henry Thatcher, after seeing a milkman making deliveries from an open bucket into which a child's filthy rag doll had accidentally fallen. By 1889, his Thatcher's Common Sense Milk Jar had become an industry standard. It was sealed with a waxed paper disc that was pressed into a groove inside the bottle's neck. The milk bottle, and the regular morning arrival of the milkman, remained a part of American life until the 1950s, when waxed paper cartons of milk began appearing in markets.

In 1990, the annual production of milk in the United States was about 148 billion lb (67.5 billion kg). This is equivalent to about 17.2 billion U.S. gallons (65.1 billion liters). About 37% of this was consumed as fluid milk and cream, about 32% was converted into various cheeses, about 17% was made into butter, and about 8% was used to make ice cream and other frozen desserts. The remainder was sold as dry milk, canned milk, and other milk products.

Types of Milk

There are many different types of milk. Some depend on the amount of milk fat present in the finished product. Others depend on the type of processing involved. Still others depend on the type of dairy cow that produced the milk.

The federal Food and Drug Administration (FDA) establishes standards for different types of milk and milk products. Some states use these standards, while others have their own standards. Prior to 1998, the federal standards required that fluid milk sold as whole milk must have no less than 3.25% milk fat, low-fat milk must have 0.5-2.0% milk fat, and skim milk must have less than 0.5% milk fat. Starting in 1998, the FDA required that milk with 2% milk fat must be labeled as "reduced-fat" because it did not meet the new definition of low-fat products as having less than 3 grams of fat per serving. Milk with 1% milk fat could still be labeled as "low-fat" because it did meet the definition. As a comparison, light cream has no less than 18% milk fat, and heavy cream has no less than 36% milk fat.

Other types of milk are based on the type of processing involved. Pasteurized milk has been heated to kill any potentially harmful bacteria. Homogenized milk has had the milk fat particles reduced in size and uniformly blended to prevent them from rising to the top in the form of cream. Vitaminfortified milks have various vitamins added. Most milk sold in markets in the United States is pasteurized, homogenized, and vitamin-fortified.

Grade A milk refers to milk produced under sufficiently sanitary conditions to permit its use as fluid milk. About 90% of the milk produced in the United States is Grade A milk. Grade B milk is produced under conditions that make it acceptable only for manufactured products such as certain cheeses, where it undergoes further processing. Certified milk is produced under exceedingly high sanitary standards and is sold at a higher price than Grade A milk.

Specialty milks include flavored milk, such as chocolate milk, which has had a flavoring syrup added. Other specialty milks include Golden Guernsey milk, which is produced by purebred Guernsey cows, and All-Jersey milk, which is produced by registered Jersey cows. Both command a premium price because of their higher milk fat content and creamier taste.

Concentrated milk products have varying degrees of water removed from fluid milk. They include, in descending order of water content, evaporated milk, condensed milk, and dry milk.

Raw Materials

The average composition of cow's milk is 87.2% water, 3.7% milk fat, 3.5% protein, 4.9% lactose, and 0.7% ash. This composition varies from cow to cow and breed to breed. For example, Jersey cows have an average of 85.6% water and 5.15% milk fat. These figures also vary by the season of the year, the animal feed content, and many other factors.

Vitamin D concentrate may be added to milk in the amount of 400 international units (IU) per quart. Most low fat and skim milk also has 2,000 IU of Vitamin A added.

Manufacturing Process

Milk is a perishable commodity. For this reason, it is usually processed locally within a few hours of being collected. In the United States, there are several hundred thousand dairy farms and several thousand milk processing plants. Some plants produce only fluid milk, while others also produce butter, cheese, and other milk products.

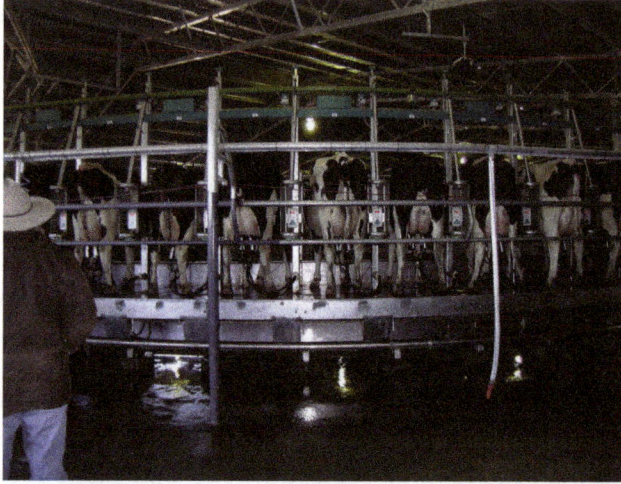

Dairy cows are milked twice a day using mechanical vacuum milking machines. The raw milk flows through stainless steel or glass pipes to a refrigerated bulk milk tank.

Collecting

- Dairy cows are milked twice a day using mechanical vacuum milking machines. The raw milk flows through stainless steel or glass pipes to a refrigerated bulk milk tank where it is cooled to about 40 °F (4.4 °C).

- A refrigerated bulk tank truck makes collections from dairy farms in the area within a few hours. Before pumping the milk from each farm's tank, the driver collects a sample and checks the flavor and temperature and records the volume.

- At the milk processing plant, the milk in the truck is weighed and is pumped into refrigerated tanks in the plant through flexible stainless steel or plastic hoses.

Separating

The cold raw milk passes through either a clarifier or a separator, which spins the milk through a series of conical disks inside an enclosure. A clarifier removes debris, some bacteria, and any sediment that may be present in the raw milk. A separator performs the same task, but also separates the heavier milk fat from the lighter milk to produce both cream and skim milk. Some processing plants use a standardizer-clarifier, which regulates the amount of milk fat content in the milk by removing only the excess fat. The excess milk fat is drawn off and processed into cream or butter.

Fortifying

Vitamins A and D may be added to the milk at this time by a peristaltic pump, which automatically dispenses the correct amount of vitamin concentrate into the flow of milk.

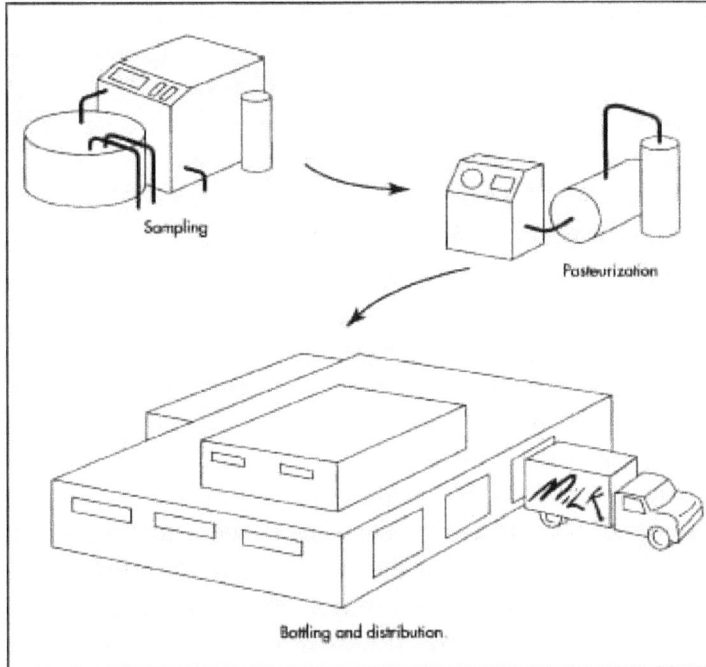

Sampling

Pasteurization

Bottling and distribution

A clarifier removes debris, some bacteria, and any sediment that may be present in the raw milk. The milk is then fortified and pasteurized.

Pasteurizing

The milk—either whole milk, skim milk, or standardized milk—is piped into a pasteurizer to kill any bacteria. There are several methods used to pasteurize milk. The most common is called the high-temperature, short-time (HTST) process in which the milk is heated as it flows through the pasteurizer continuously. Whole milk, skim milk, and standardized milk must be heated to 161 °F (72 °C) for 15 seconds. Other milk products have different time and temperature requirements. The hot milk passes through a long pipe whose length and diameter are sized so that it takes the liquid exactly 15 seconds to pass from one end to the other. A temperature sensor at the end of the pipe diverts the milk back to the inlet for reprocessing if the temperature has fallen below the required standard.

Homogenizing

- Most milk is homogenized to reduce the size of the remaining milk fat particles. This prevents the milk fat from separating and floating to the surface as cream. It also ensures that the milk fat will be evenly distributed through the milk. The hot milk from the pasteurizer is pressurized to 2,500-3,000 psi (17,200-20,700 kPa) by a multiple-cylinder piston pump and is forced through very small passages in an adjustable valve. The shearing effect of being forced through the tiny openings breaks down the fat particles into the proper size.

- The milk is then quickly cooled to 40 °F (4.4 °C) to avoid harming its taste.

Packaging

- The milk is pumped into coated paper cartons or plastic bottles and is sealed. In the United States most milk destined for retail sale in grocery stores is packaged in one-gallon (3.8-liter) plastic bottles. The bottles or cartons are stamped with a "sell by" date to ensure that the retailers do not allow the milk to stay on their shelves longer than it can be safely stored.

- The milk cartons or bottles are placed in protective shipping containers and kept refrigerated. They are shipped to distribution warehouses in refrigerated trailers and then on to the individual markets, where they are kept in refrigerated display cases.

Cleaning

To ensure sanitary conditions, the inner surfaces of the process equipment and piping system are cleaned once a day. Almost all the equipment and piping used in the processing plant and on the farm are made from stainless steel. Highly automated clean-in-place systems are incorporated into this equipment that allows solvents to be run through the system and then flushed clean. This is done at a time between the normal influx of milk from the farms.

CEREAL PROCESSING

Cereal processing is the treatment of cereals and other plants to prepare their starch for human food, animal feed, or industrial use.

Cereals, or grains, are members of the grass family cultivated primarily for their starchy seeds (technically, dry fruits). Wheat, rice, corn (maize), rye, oats, barley, sorghum, and some of the millets are common cereals; their composition is shown in the table.

Table: Nutrient composition of selected raw cereal grains (per 100 grams).

Cereal grain	Energy (kcal)	Water (g)	Carbohydrate (g)	Protein (g)	Fat (g)	Minerals (g)
Barley (pearled)	352	10.09	77.72	9.91	1.16	1.11
Corn (field)	365	10.37	74.26	9.42	4.74	1.20
Millet	378	8.67	72.85	11.02	4.22	3.25
Oats (oatmeal)	384	8.80	67.00	16.00	6.30	1.90

Cereal grain	Energy (kcal)	Water (g)	Carbohydrate (g)	Protein (g)	Fat (g)	Minerals (g)
Rice (brown; long-grain)	370	10.37	77.24	7.94	2.92	1.53
Rye	335	10.95	69.76	14.76	2.50	2.02
Sorghum	339	9.20	74.63	11.30	3.30	1.57
Wheat (hard red winter)	327	13.10	71.18	12.61	1.54	1.57

Starch, a carbohydrate stored in most plants, is a major constituent of the average human diet, providing a low-cost energy source with good keeping qualities. Cereals are high in starch, which may be used in pure or flour form. Starches are also obtained from such root sources as potatoes and from the pith of tropical palm trees. Various starches are used commercially in food processing and in the manufacture of laundering preparations, paper, textiles, adhesives, explosives, and cosmetics.

Cereal Processing and Utilization

Milling

Cereal processing is complex. The principal procedure is milling—that is, the grinding of the grain so that it can be easily cooked and rendered into an attractive foodstuff. Cereals usually are not eaten raw, but different kinds of milling (dry and wet) are employed, depending on the cereal itself and on the eating customs of the consumer. Wheat may be crushed with grinding stones or similar devices or by modern automated systems employing steel cylinders, followed by air purification and numerous sievings to separate the endosperm from the outer coverings and the germ.

Corn is often milled by wet processes, but dry milling is also practiced, especially in the developing countries. Corn, with its high germ content, is inclined to respire more during storage and, unless precautions are taken, may increase in temperature during incorrect storage. Most other cereals are ground in the dry state. Some cereal grains are polished, removing most of the bran and germ and leaving the endosperm.

Uses

Human Food

Cereals are used for both human and animal food and as an industrial raw material. Although milled white flour is largely used for bread production, especially in industrialized countries, the grain may be converted to food in other ways.

Animal Food

The principal cereals used as components of animal feeds are wheat and such wheat by-products as the outer coverings separated in the preparation of white flour (bran and the more floury middlings), corn, barley, sorghum, rye, and oats. These are supplemented by protein foods and green fodders.

Animal foods require proper balance between the cereals (carbohydrates) and the more proteinous foods, and they must also contain suitable amounts of necessary minerals, vitamins, and other nutrients. The compounded ration for a milking cow generally contains about 50–80 percent cereals, consisting of wheat by-products, flaked or ground corn, barley, sorghum, wheat, and oats. Requirements for most balanced rations for pigs and poultry are similar. Corn is especially useful in high-energy feeds either as meal or as the flaked and partly gelatinized product; barley is desirable for fattening, and oats help provide a better balanced cereal for livestock. Without cereals for use in farm animal foods, the available supply of the animal protein required in the human diet would be greatly reduced.

Industrial Uses

The relatively minor use of cereals in nonfood products includes the cellulose in the straw of cereals by the paper industry, flour for manufacturing sticking pastes and industrial alcohol, and wheat gluten for core binders in the casting of metal. Rice chaff is often used as fuel in Asia.

Wheat: Varieties and Characteristics

The three principal types of wheat used in modern food production are Triticum vulgare (or aestivum), T. durum, and T. compactum. T. vulgare provides the bulk of the wheat used to produce flour for bread making and for cakes and biscuits (cookies). It can be grown under a wide range of climatic conditions and soils. Although the yield varies with climate and other factors, it is cultivated from the southernmost regions of America almost to the Arctic and at elevations from sea level to over 10,000 feet. T. durum, longer and narrower in shape than T. vulgare, is mainly ground into semolina (purified middlings) instead of flour. Durum semolina is generally the best type for the production of pasta foods. T. compactum is more suitable for confectionery and biscuits than for other purposes.

The wheat grain, the raw material of flour production and the seed planted to produce new plants, consists of three major portions: (1) the embryo or germ (including its sheaf, the scutellum) that produces the new plant, (2) the starchy endosperm, which serves as food for the germinating seed and forms the raw material of flour manufacture, and (3) various covering layers protecting the grain. Although proportions vary, other cereal grains follow the same general pattern. Average wheat grain composition is approximately 85 percent endosperm, 13 percent husk, and 2 percent embryo.

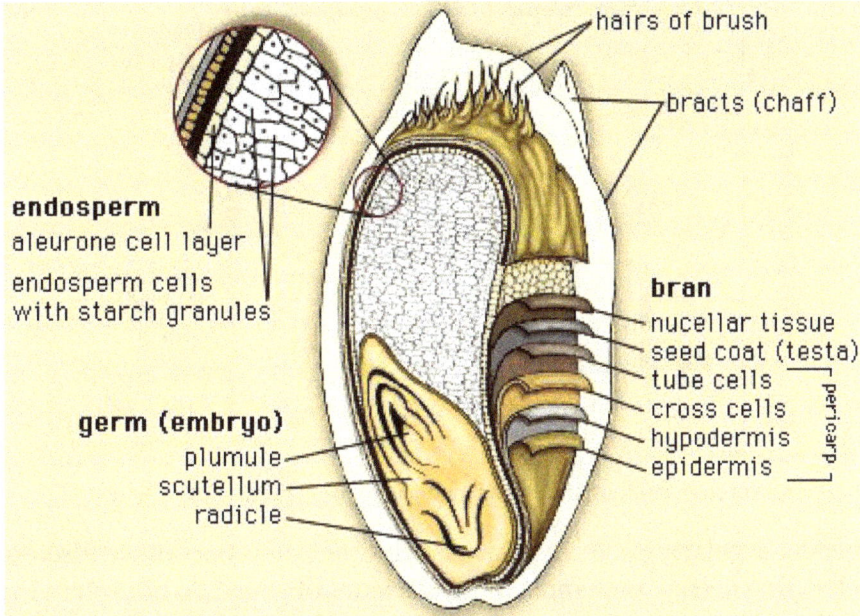

The outer layers and internal structures of a kernel of wheat.

Characteristic variations of the different types of wheat are important agricultural considerations. Hard wheats include the strong wheats of Canada (Manitoba) and the similar hard red spring (HRS) wheats of the United States. They yield excellent bread-making flour because of their high quantity of protein (approximately 12–15 percent), mainly in the form of gluten. Soft wheats, the major wheats grown in the United Kingdom, most of Europe, and Australia, result in flour producing less attractive bread than that achieved from strong wheats. The loaves are generally smaller, and the crumb has a less pleasing structure. Soft wheats, however, possess excellent characteristics for the production of flour used in cake and biscuit manufacture.

Wheats intermediate in character include the hard red winter (HRW) wheats of the central United States and wheat from Argentina. There are important differences between spring and winter varieties. Spring wheats, planted in the early spring, grow quickly and are normally harvested in late summer or early autumn. Winter wheats are planted in the autumn and harvested in late spring or early summer. Both spring and winter wheats are grown in different regions of the United States and Russia. Winter varieties can be grown only where the winters are sufficiently mild. Where winters are severe, as in Canada, spring types are usually cultivated, and the preferred varieties mature early, allowing harvesting before frost.

In baking and confectionery, the terms strong and weak indicate flour from hard and soft wheats, respectively. The term strength is used to describe the type of flour, strong flours being preferred for bread manufacture and weak flours for cakes and biscuits. Strong flours are high in protein content, and their gluten has a pleasing elasticity; weak flours are low in protein, and their weak, flowy gluten produces a soft, flowy dough.

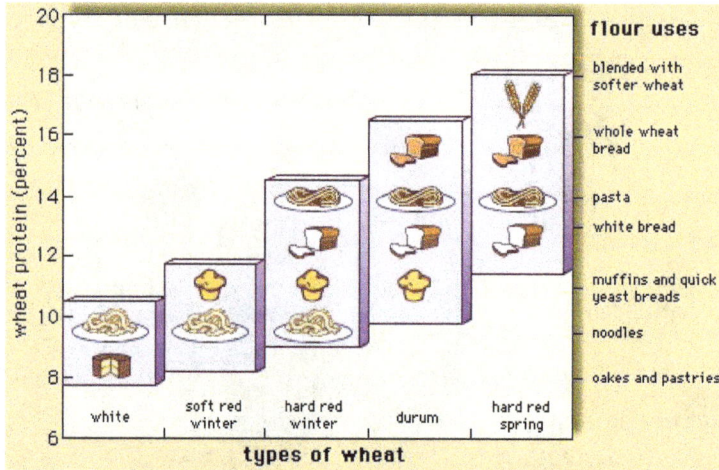

The protein content and major food uses of certain varieties of wheat.

Wheat breeders regularly produce new varieties, not only to combat disease but also to satisfy changing market demands. Many varieties of wheat do not retain their popularity, and often those popular in one decade are replaced in the next. New varieties of barley have also been developed, but there have been few varieties of rice.

Wheat Flour

The milling of wheat into flour for the production of bread, cakes, biscuits, and other edible products is a huge industry. Cereal grains are complex, consisting of many distinctive parts. The objective of milling is separation of the floury edible endosperm from the various branny outer coverings and elimination of the germ, or embryo. Because wheats vary in chemical composition, flour composition also varies.

Although some important changes have occurred in flour milling, basic milling procedure during the past 100 years has employed the gradual reduction process.

Milling

In modern milling considerable attention is given to preliminary screening and cleaning of the wheat or blend of wheats to exclude foreign seed and other impurities. The wheat is dampened and washed if it is too dry for subsequent efficient grinding, or if it is too damp it is gently dried to avoid damaging the physical state of the protein present, mainly in the form of the elastic substance gluten.

The first step in grinding for the gradual reduction process is performed between steel cylinders, with grooved surfaces, working at differential speeds. The wheat is directed between the first "break," or set of rolls, and is partially torn open. There is little actual grinding at this stage. The "chop," the resulting product leaving the rolls, is sieved, and three main separations are made: some of the endosperm, reduced to flour called "first break flour"; a fair amount of the coarse nodules of floury substances from the endosperm, called semolina; and relatively large pieces of the grain with much of the endosperm still adhering to the branny outsides. These largish portions of the wheat are fed to the second break roll. The broad objective of this gradual reduction process is the release, by means of the various sets of break rolls, of inner endosperm of the grain, in the form of semolina, in amounts sufficient that the various semolinas from four or five break rolls can be separated by suitable sieving and the branny impurities can be removed by air purifiers and other devices. The cleaned semolinas are reduced to fine flour by grinding between smooth steel rolls, called reduction rolls. The flour produced in the reduction rolls is then sieved out. There are usually four or five more reduction rolls and some "scratch" rolls to scrape the last particles of flour from branny stocks. Since the various sieving and purification processes free more and more endosperm in the form of flour, flour is obtained from a whole series of processing operations. The flour is sieved out after each reduction roll, but no attempt is made to reduce to flour all the semolina going to a particular reduction roll. Some of the endosperm remains in the form of finer semolina and is again fed to another reduction roll. Each reduction roll tends to reduce more of the semolina to flour and to flatten bran particles and thus facilitate the sieving out of the branny fractions. The sieving plant generally employs machines called plan-sifters, and the air purifiers also produce a whole series of floury stocks.

Modern flour processing consists of a complicated series of rolls, sieves, and purifiers. Approximately 72 percent of the grain finally enters the flour sack.

The sacked flour may consist of 20 or more streams of flour of various states of purity and freedom from branny specks. By selection of the various flour streams it is possible to make flour of various grades. Improvements in milling techniques, use of newer types of grinding machinery in the milling system, speeding up of rolls, and improved skills have all resulted in flour produced by employing the fundamentals of the gradual reduction process but with simplified and shorter milling systems. Much less roll surface is now required than was needed as recently as the 1940s.

The purest flour, selected from the purest flour streams released in the mill, is often called patent flour. It has very low mineral (or ash) content and is remarkably free from traces of branny specks and other impurities. The bulk of the approximately 72 percent released is suited to most bread-making purposes, but special varieties are needed for some confectionery purposes. These varieties may have to be especially fine for production of specialized cakes, called high-ratio cakes, that are especially light and have good keeping qualities.

In many countries the flour for bread production is submitted to chemical treatments to improve the baking quality.

In modern processing, regrinding of the flour and subsequent separation into divisions by air treatment has enabled the processors to manufacture flour of varying protein content from any one wheat or grist of wheats.

Composition and Grade

Flour consists of moisture, proteins (mainly in gluten form), a small proportion of fat or lipids, carbohydrates (mainly starch, with a small amount of sugar), a trace of fibre, mineral matter (higher amounts in whole meal), and various vitamins. Composition varies among the types of flour, semolinas, middlings, and bran.

Protein Content

For bread making it is usually advantageous to have the highest protein content possible (depending on the nature of the wheat used), but for most other baked products, such as cookies (sweet biscuits) and cakes, high protein content is rarely required. Gluten can easily be washed out of flour by allowing a dough made of the flour and water to stand in water a short time, followed by careful washing of the dough in a gentle stream of water, removing the starch and leaving the gluten. For good bread-making characteristics, the gluten should be semi-elastic, not too stiff and unyielding but not soft and flowy, although a flowy quality is required for biscuit manufacture.

The gluten, always containing a small amount of adhering starch, is essentially hydrated protein. With careful drying it will retain its elasticity when again mixed with water and can be used to increase the protein content of specialized high-protein breads.

Sometimes locally grown wheat, often low in protein, may be the only type available for flour for bread making. This situation exists in parts of France, Australia, and South Africa. The use of modern procedures and adjustment of baking techniques, however, allow production of satisfactory bread. In the United Kingdom, millers prefer a blend of wheat, much of it imported, but modern baking procedures have allowed incorporation of a larger proportion of the weak English wheat than was previously feasible.

Treatment of Flour

Use of "improvers," or oxidizing substances, enhances the baking quality of flour, allowing production of better and larger loaves. Relatively small amounts are required, generally a few parts per million. Although such improvers and the bleaching agents used to rectify excessive yellowness in flour are permitted in most countries, the processes are not universal. Improvers include bromates, chlorine dioxide (in gaseous form), and azodicarbonamide. The most popular bleacher used is benzoyl peroxide.

Grade

The grade of flour is based on freedom from branny particles. Chemical testing methods are employed to check general quality and particularly grade and purity. Since the ash (mineral content) of the pure branny coverings of the wheat grain is much greater than that of the pure endosperm, considerable emphasis is placed on use of the ash test to determine grade. Bakers will generally pay higher prices for pure flour of low ash content, as the flour is brighter and lighter in colour. Darker flours may have ash content of 0.7 to 0.8 percent or higher.

A widely employed modern method for testing flour colour is based on the reflectance of light from the flour in paste form. This method requires less than a minute; the indirect ash test requires approximately one to two hours.

Nonwheat Cereals

Barley

Most of the barley grown in the world is used for animal feed, but a special pure barley is the source of malt for beer production. Barley is also used in the manufacture of vinegar, malt extract, some milk-type beverages, and certain breakfast foods. In addition, in flaked form it is employed in some sections of the brewing industry, and pearl barley (skins removed by emery friction) is used in various cooked foods.

Barley can be cultivated on poorer soil and at lower temperatures than wheat. An important characteristic in barley is "winter-hardiness," which involves the ability to modify or withstand many types of stresses, particularly that of frost. However, barley is subject to many of the diseases and pests that affect wheat.

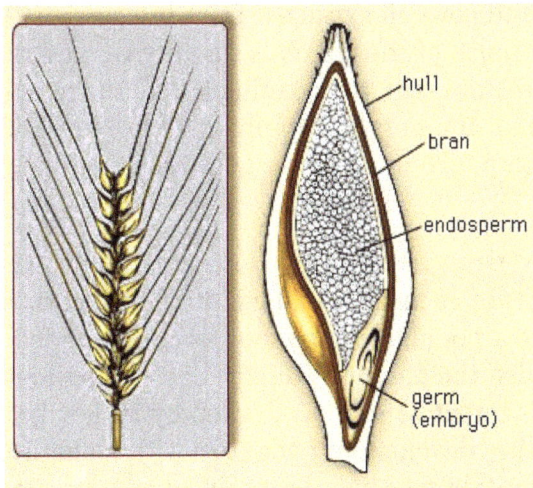

(Left) The barley spike, with rows of barley florets.
(Right) Cross section of the barleycorn.

The use of barley in animal feed is increasing; it has been a basic ingredient of pig foods for years and is increasingly used for cattle feed. Its use in poultry foods has decreased because it has a lower starch equivalent when compared with wheat or corn and thus provides a lower-energy ration, unsuitable in modern poultry production. Barley vitamin content is similar to that of wheat.

Corn

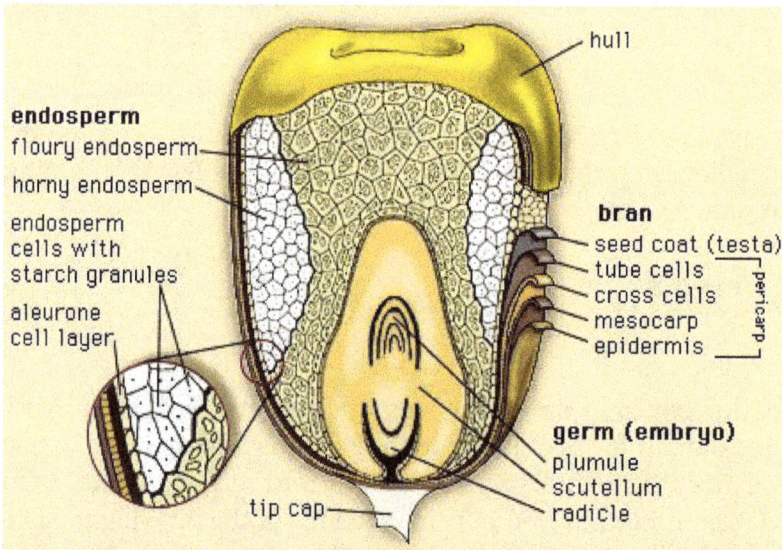

The outer layers and internal structures of a kernel of corn.

Corn, or maize, a cereal cultivated in most warm areas of the world, has many varieties. The United States, the principal producer of corn, cultivates two main commercial types, Zea indurata (flint corn) and Z. indentata (dent corn). The plant grows to a height of about three metres or more. The corn kernel is large for a cereal, with a high embryo content, and corn oil extracted from the germ is commercially valuable. The microscopic appearance of the starch is distinctive, and the principal protein in ordinary corn is the prolamin zein, constituting half of the total protein. On hydrolysis zein yields only very small amounts of tryptophan or lysine, making it low in biological value. The proteins of corn, like those of most cereals other than wheat, do not provide an elastic gluten.

Much of the corn is wet-processed to produce corn flour, widely used in cooking. Corn, dry-milled as grits or as meal or turned into flaked corn with some of its starch partially gelatinized, is a popular component in compounded animal feedstuffs. In dry-milled form it is also the basis of human food throughout large areas of Africa and South America. Its nutritive value is limited by its low lysine content. Much recent research has involved development of a corn with higher lysine content. Mutants have been produced containing much less zein but possessing protein with higher than normal lysine and tryptophan contents, sometimes increased as high as 50 percent. These corns, called Opaque-2 and Floury-2, possess certain drawbacks. They are generally lower in yield than dent hybrids, are subject to more kernel damage when combine-harvested, and may be more difficult to process. Nevertheless, these new hybrid corns are expected to become widely cultivated, and the principles involved in their production may also be applied to sorghum, wheat, and rice. Corn is popular for use in breakfast foods.

Sorghum

Sorghum, also called milo, is of smaller size than corn but is generally the same type of cereal, with similar appearance. Its numerous types are mainly used for animal feeding. It is grown extensively in the United States, Pakistan, central India, Africa, and China. In the sorghum endosperm, the proteins soluble in hot 60 percent alcohol, called kafirin, constitute the major portion of the protein. Milo germ oil is similar to corn germ oil; its major fatty acids are palmitic, stearic, and particularly oleic and linoleic. Milo is commercially graded in the United States. In waxy varieties the starch is principally in the form of amylopectin, with very little amylose. Such starches possess special viscosity characteristics.

Oats

Oats belong to the botanical genus Avena, which includes a large number of types, the principal being A. sativa, A. sterilisand A. strigosa. Oats are widely grown in most countries but are not suitable for Mediterranean climates. Oats are frequently grown on farms as feed for the farm's livestock. They are well balanced chemically, with fairly high fat content, and are particularly suitable for feeding horses and sheep.

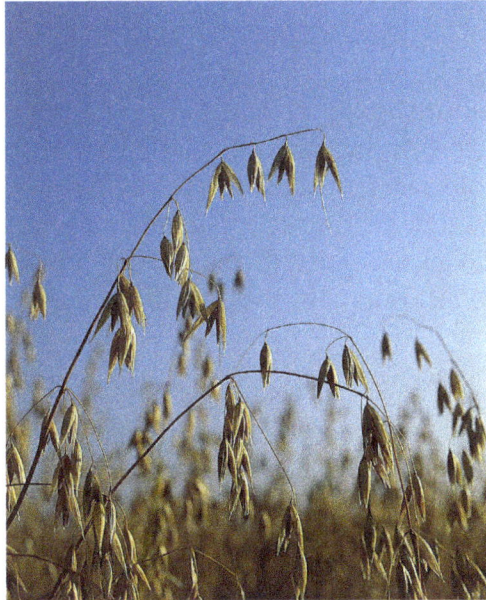

Mature oats (Avena sativa).

Although a large portion of the world's oat production is used for animal feed, oatmeal is a popular human food in many countries. Thin-skinned grains, fairly rich in protein and not too starchy, are selected. Preliminary cleaning is essential for human consumption. The oats are then kilned (roasted). Thin-husked oats yield 60 percent oatmeal; varieties with thick husks yield only 50 percent.

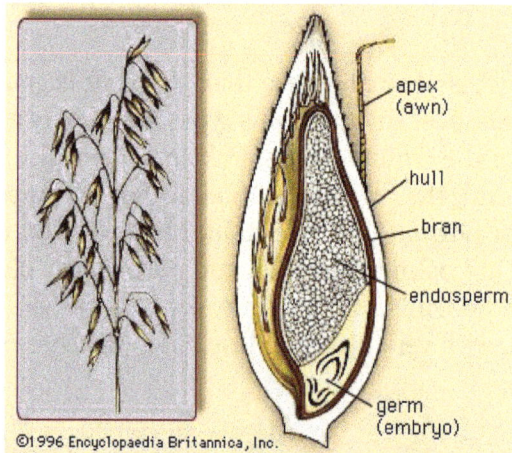

(Left) The oat panicle, bearing multiple oat florets.
(Right) Cross section of the oat grain.

Rapid development of rancidity is a serious problem in oats and oat products. The free fatty acid content must be controlled because formation of these acids tends to produce a soapy taste resulting from the activity of the enzyme lipase. A few minutes of steam treatment normally destroys the lipase activity in the grain.

Rye

Rye, which has been known for some 2,000 years, ranks second to wheat as a bread flour. The principal rye producers are Russia, Poland, Belarus, Germany, and Ukraine. The popularity of true rye bread is decreasing, and a similar bread, retaining some of the original characteristics, is now made from a rye and wheat blend. The protein of European rye tends to be low and does not yield gluten in the same way as does wheat. Rye bread, closer-grained and heavier than wheat bread, is aerated by the use of a leaven (sourdough) rather than yeast. The grain is susceptible to attack by the parasitic fungus ergot (Claviceps purpurea).

Rice

Cultivated rice is known botanically as Oryza sativa, only one of some 25 species comprising the genus Oryza. The importance of this cereal to certain parts of the world may be seen from the fact that in Sanskrit there exists, besides the usual word for rice, another term signifying "sustainer of the human race." Rice is the staple food for millions in Southeast Asia, almost equal to wheat in importance among the world's cereal crops.

Cultivation

More than 90 percent of the world's rice is grown in Asia, principally in China, India, Indonesia, and Bangladesh, with smaller amounts grown in Japan, Pakistan, and various Southeast Asian nations. Rice is also cultivated in parts of Europe, in North and South America, and in Australia. The bulk of the rice cultivated in Asia is grown under water in flooded fields. Successful production depends on adequate irrigation, including construction of dams and waterwheels, and on the quality of the soil. Long periods of sunshine are essential. Rice yields vary considerably, ranging from 700 to 4,000 kilograms per hectare (600 to 3,500 pounds per acre). Adequate irrigation, which means inundation of the fields to a depth of several inches during the greater part of the growing season, is a basic requirement for productive land use.

Dryland paddy production, with harvesting by modern mechanical means, is limited to a few areas, and it produces only a fraction of the total world crop.

As with other cereals, weeds, especially wild red rice, are a constant problem. The commonest pests include plant bugs, stem borers, worms, and grasshoppers. The crop, often harvested with a sickle, is frequently dried in earth or concrete pits. Threshing is often carried out by trampling or with crude implements. Only in a few rice-growing regions are more modern procedures used in harvesting.

Manpower requirements for crops vary enormously, but over 400 man-hours per acre are required in smallholdings in Asia, where labour is cheap.

In Asia the paddy is cultivated in three main types of soil, including clays with a firm bottom within a few inches of the surface; silts and soft clays with soft bottoms becoming hard on drying; and peats and "mucks" containing peat, provided the depth of the peat is not excessive. Fields must be drained and dried before harvesting. When combine harvesters or binder threshers are employed, the grain must be dried to about 14 percent moisture so that no deterioration takes place in storage. When reaper binders are used, the crop is "shocked" in certain ways so that the grain is protected from rain.

Milling

Milling methods used in most of Asia are primitive, but large mills operate in Japan and some other areas. Hulling of the paddy is usually accomplished by pestle and mortar worked by hand, foot, or water power. Improvements are slowly taking place. The yield of milled rice is dependent on the size and shape of the grain, the degree of ripeness, and the extent of exposure to the sun. Some large mills, handling 500 to 1,000 tons of paddy daily, have specialized hulling plants with consequent smaller losses from broken grain. They generally employ modern milling techniques and rely on controlled drying plants instead of on sun drying.

The weight of the husk is about 20 percent of the weight of the paddy, and there are losses of about 5 percent from dirt, dead grains, and other impurities. Approximately 74 percent of the paddy is available as rice and rice by-products. The yield from milling and subsequent emery polishings includes about 50 percent whole rice, 17 percent broken rice, 10 percent bran, and 3 percent meal. Rice grains have a series of thin coats that can be removed or partially removed in the process of pearling and whitening.

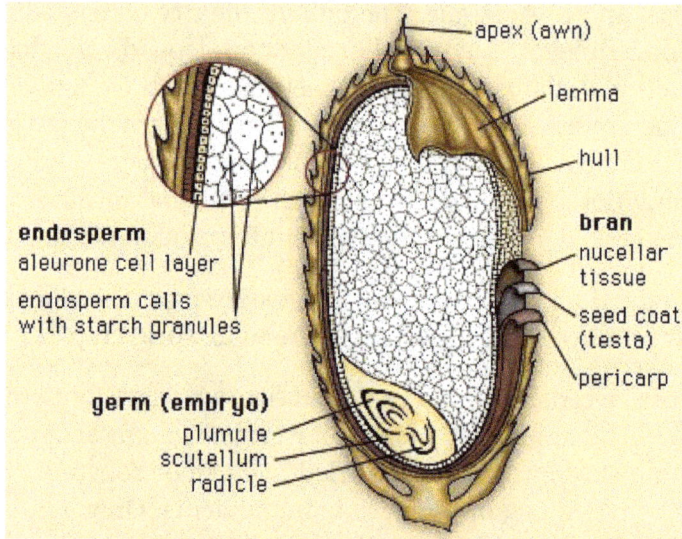

The outer layers and internal structures of a rice grain.

About 60 percent of the Indian rice is parboiled. In the parboiling process the paddy is steeped in hot water, subjected to low-pressure steam heating, then dried and milled as usual. Parboiling makes more rice available from the paddy, and more nutrients (largely vitamin B1) are transferred from the outer coverings to the endosperm, improving the nutritive value of the finished product. Parboiled rice may contain two to four times as much thiamine (vitamin B1) and niacin as milled raw rice, and losses in cooking may also be reduced.

Alcoholic drinks, such as sake in Japan and wang-tsin in China, are made from rice with the aid of fungi. The hull or husk of paddy, of little value as animal feed because of a high silicon content that is harmful to digestive and respiratory organs, is used mainly as fuel.

Nutritive Value

The lysine content of rice is low. As rice is not a complete food, and the majority of Asians live largely on rice, it is important that loss of nutrients in processing and cooking should be minimal. Lightly milled rice has about 0.7 milligram of vitamin B1 per 1,000 nonfatty calories, and the more costly highly milled product has only 0.18 milligram of B1 on the same basis. For adequate nutrition, vitamin B1 in the daily diet on this basis should be 0.5–0.6 milligram. The amount of fat-soluble vitamins in rice is negligible.

In some countries rice is enriched by addition of synthetic vitamins. According to U.S. standards for enriched rice, each pound must contain 2–4 milligrams of thiamine, 1.2–2.4 milligrams of riboflavin, 16–32 milligrams of niacin, and 13–26 milligrams of iron. In enriched rice the loss of water-soluble vitamins in cooking is much reduced because enrichment is applied to about 1 grain in 200, and these enriched grains are protected by a collodion covering. In ordinary rice, especially when open cookers are employed or excessive water is used, nutrient losses can be high.

Millet

This term is applied to a variety of small seeds originally cultivated by the ancient Egyptians, and Romans and still part of the human diet in China, Japan, and India, though in Western countries it is used mainly for birdseed. The genus is termed Panicum. The small seed is normally about two millimetres long and nearly two millimetres broad. The term proso is one of several alternative names. Japanese barnyard millet is a well-known variety.

Other Starch-yielding Plants

Cassava

Cassava, often called manioc, is not a cereal but a tuber; however, it replaces cereals in certain countries, supplying the carbohydrate content of the diet. The botanical name is Manihot esculenta, and the plant is native to South America, especially Brazil. It is now grown in Indonesia, Malaysia, the Philippines, Thailand, and parts of Africa. A valuable source of starch, cassava is familiar in many developed countries in a granular form known as tapioca.

Easily cultivated and curiously immune to most food-crop pests, cassava is a staple crop in several areas of Latin America. The actual tubers may weigh up to 14 kilograms (30 pounds). Some tubers may be bitter and contain dangerously large amounts of prussic acid.

Dry milling of cassava is rarely practiced because it yields a product inferior to wet-processed starch in which the tubers are crushed or rasped with water and the starch is permitted to settle. Wet starch is dried to a point where it can be crumbled by pressing it through metal plates (or sieves). This crumbled material is subjected to a rotary motion, sometimes carried out on canvas cloth fastened to cradle-shaped frames. Another method is to tumble the material in revolving steam-jacketed cylinders so that the particles assume a round pellet form and are partially gelatinized as they dry. Sun drying is employed in both homes and small mills.

Many tapioca factories and mills are equipped with modern raspers, special shaking or rotating sieves, and settling tanks of various types; but some fermentation takes place, and small rural mills can often be identified by the smell of butyric acid. In larger mills, centrifuges are replacing the settling tanks.

For its various industrial uses, the tuber usually goes under its alternative name, manioc. It is used in the textile industries, explosives manufacture, leather tanning, and production of glues and dextrins and alcohol.

Fresh cassava leaves are rich in protein, calcium, and vitamins A and C. Their prussic acid level must be reduced to safe limits by boiling; the duration of boiling depends on the variety of the leaves. Cassava leaves are a popular vegetable in Africa, and the tuber also is used in meal for animal feed.

Soybean

Soybean (Glycine max) is not a cereal but a legume; because of its widespread use in the baking industry, it may appropriately be dealt with here. Soybean provides protein of high biological value. Although Asia is its original source, the United States became the major world producer in the late 20th century.

The valuable oil of the soybean, widely used in industry, is extracted either by solvents or by expellers. The amino acid distribution of soy protein is more like that found in animal protein than the protein from most vegetable sources; for example, lysine comprises about 5.4 percent. The oil content includes useful amounts of phosphorus; the phosphatide content of soy flour is about 2 percent and is a mixture of lecithin and cephalin. The low carbohydrate content exists mainly as sugars. The table shows the amino acid composition of soy protein.

Amino acid composition of soy protein (calculated to 16 percent nitrogen)	
Amino acid	Percent
Arginine	5.8
Histidine	2.3
Lysine	5.4
Tyrosine	4.1
Tryptophan	1.2
Phenylalanine	5.7
Cystine	0.9
Methionine	2.0
Threonine	4.0
Leucine	6.6
Isoleucine	4.7
Valine	4.2
Glutamic acid	21.0
Aspartic acid	8.8

Although soybeans are a good source of thiamine, much of this may be lost in processing. Average vitamin contents of soybean (as micrograms per gram) are as follows: thiamine 12, riboflavin 3.5, nicotinic acid 23, pyridoxine 8, pantothenic acid 15, and biotin 0.7.

The bulk of the soybean produced in the United States is used for animal feed; the Asian crop goes principally for human diet.

Soybean milk is produced and used in the fresh state in China and as a condensed milk in Japan. In both of these preparations, certain antinutritive factors (antitrypsin and soyin) are largely removed. In the Western world most soy products are treated chemically or by heat to remove these antinutritive factors along with the unpopular beany taste. Such processing affects the enzymatic activity in the milk.

Soybean is milled to produce soy flour. The flour is often used in a proportion of less than 1 percent in bakery operations. It stiffens doughs and helps to maintain crumb softness. Unprocessed soy flour, because of its lipoxidase enzyme system, is employed with high-speed mixing to bleach the flour in a dough.

In addition to their use in bread, soy products are used in confectionery, biscuits, macaroni, infant and invalid foods, ice cream, chocolate, sausages, sauces, lemon curd, mayonnaise, meat and fish pastes, certain diabetic foods, and in such nonfood products as paint, paper, textiles, and plastics.

A recent development is the isolation of the soybean proteins for use as emulsifiers and binders in meat products and substitutes. Enzyme-modified proteins provide useful egg-albumen supplement for whipped products.

Buckwheat

Botanically, buckwheat is not a cereal but the fruit of Fagopyrum esculentum. Its name is probably derived from its resemblance to beechnut. Believed to have originated in China, the plant grows to a height of about one metre and thrives best in cool, moist climates, although it does not easily tolerate frost. It can be grown on a wide range of soils, and a crop can be obtained within 10–12 weeks of sowing. The seed is dark brown in colour and often triangular in shape. It contains about 60 percent carbohydrate, 10 percent protein, and 15 percent fibre. A white flour can be obtained from the seeds (buckwheat cakes and pancakes are popular in certain areas), and buckwheat meal is also used in animal feed. The whole seed may be fed to poultry and game birds. There is some medical interest in buckwheat as a source of rutin, possibly effective in treatment of increased capillary fragility associated with hypertension in humans.

Starch Products

Commercial Starches

Starch has been used for many centuries. An Egyptian papyrus paper dating from 3500 BCE was apparently treated with a starch adhesive. The major starch sources are tubers, such as potatoes and cassava, and cereals. Current starch production is considerable. Among the major producing areas, the European countries use both domestic wheat and potatoes and imported corn as the raw material; the United States uses corn and such similar cereals as sorghum; and in South America the cassava plant is the major raw material.

Separated from tubers and cereals, starch is used for conversion into various sugars, and half of the world's separated starch is processed into glucose. Starch is also processed for use in adhesives manufacture. In the food industry starch is used as a thickener in the preparation of cornstarch puddings, custards, sauces, cream soups, and gravies. Starch from tubers and cereals provides the carbohydrate of the human diet.

Large quantities of starch and its derivatives are used in the paper and textile industries.

Starch from Tubers

In Germany, the Netherlands, Poland, and a number of other countries, the extraction of the starch from potatoes (sometimes called farina) is a major industry. Some factories produce over 300 tons daily. Processing involves continuous and automatic cleaning of the potatoes, thorough disintegration in raspers or hammer mills, and separation of the fibres from the pulp by centrifugal (rotary) sieves. The resulting starch "milk" contains starch in suspension and soluble potato solids in solution. The starch is separated and washed free from the solubles, the water is removed by centrifugal action, and the damp starch is dried. The flash type of dryer, using hot air, is widely employed for starches derived from both tubers and cereals. Sulfurous acid is generally introduced into the process to prevent the development of various microorganisms.

Potato flour is also produced in Germany and other countries, slices of cleaned potatoes being dried, ground, and sieved. In Germany a "potato sago" is produced. The starch cake obtained from the potatoes is crumbled to produce reasonably uniform-size particles that are rounded by tumbling or similar operations, heated to gelatinize the outside layers of the starch, and then dried.

Potatoes were employed in baking to make the barm, or leaven, before compressed distiller's yeast was available, and they have also been used to supplement limited supplies of wheat flour. The potatoes are cleaned, boiled until soft but not mushy, and mixed, in a proportion of 2 to 3 percent, in the dough.

Modern, ready-to-use, dried and powdered mashed potatoes are popular consumer products.

Cassava and tapioca starches are sometimes partially gelatinized by vacuum drying. Protein impurities are low in commercial starches of potato, sago, and tapioca but as high as 0.2 percent in wheat starch and higher in corn flour.

Cornstarch

Corn is wet-milled to produce corn flour, or cornstarch, desirable for cooking because it forms a paste that sets with a "short" texture and separates from molds more cleanly than do the gels produced by such starches as potato, tapioca, and arrowroot, which are "long," or elastic. In wet milling, the grains are first dry-cleaned so that other cereals and some of the impurities are removed, then steeped in warm water containing sulfur dioxide. This process softens the grains, and the outer skin and the germ are rendered removable. The corn is coarsely ground in "degerminating mills," and the slurry is further wet-ground and sieved to remove all the germ and complete the separation of the starch.

The germ, rich in oil, is eventually dried, and the oil is expelled by pressure, providing an excellent edible oil for culinary use, often replacing olive oil. Corn oil is used for salad oil, margarine, and shortening and for such nonfood items as soap.

The pure starch, held in suspension, was formerly collected by gravity as it flowed down tables, but in modern practice the starch suspension is thickened by the elimination of water by means of machines, and the starch is finally separated by the use of centrifuges. The starch is readily dried without gelatinization taking place.

There is a regular demand for a good grade of corn flour, or cornstarch. Roller-milled corn is still produced for human consumption in Africa and elsewhere. In the United States some corn grits are used by brewers, but the bulk of the corn grown is used for animal feed as meal, grits, or in partially gelatinized flake form.

Rice Starch

Rice starch, largely used in laundry work, is normally prepared from broken white rice. The broken grains are steeped for several hours in a caustic soda solution, and the alkali is finally washed away with water. The softened grains are ground with more caustic soda solution, and the resulting mass is settled or submitted to centrifugation in a drum. The starch layer is agitated with water (often with 0.25 percent formaldehyde solution added), and the resulting starch liquor is dewatered, washed on a continuous rotary vacuum filter, resuspended in water, and finally dewatered in a perforated basket centrifuge to about 35 percent moisture. In modern processing it is usual to roll out a thick layer of moist starch, which is then slowly dried and falls to pieces as crystals.

Starch Composition

Starch consists of two components: amylose and amylopectin. The relative proportion of these two components varies, and they react differently to enzymatic attack. The enzyme β-amylase (maltogenic) attacks the straight chain amylose but is unable to attack most of the branch chain amylopectin. If only β-amylase is present, maltose is produced, together with a residue of the amylopectin portion, or dextrin of high molecular weight. When α-amylase (dextrinogenic) attacks starch, gummy dextrins of low molecular weight are formed and can produce a sticky crumb in bread.

In bread making there is only limited time for such enzymatic attacks on the starch, and only the "attackable" or "damaged" granules can produce the fermentable sugar for the dough. The β-amylase has little effect on viscosity. The viscosity of gelatinized starch is markedly reduced by α-amylase, however, and is therefore valuable in syrup and dextrose manufacture.

The gelatinization of starch that occurs in hot water is an important characteristic, and the viscous pastes formed are influenced by the treatment the starch has received in its preliminary separation from the cereal or tuber. Chemicals affect degree and speed of gelatinization and the nature and viscosity of the pastes formed.

In certain cereals, particularly in special corns, the starch consists almost entirely of amylopectin, and the term "waxy" is applied to such cereals. They are useful for their unusual physical properties and viscosities. They possess outstanding paste clarity, high water-binding capacity, and resistance to gel formation and retrogradation; they are helpful in production of salad dressings, sauces, and pie fillings and in some canned goods; they are useful because of resistance to irreversible gel formation and syneresis on freezing and especially for many products stored in the frozen state.

Processing

The carbohydrate starch is rarely consumed in the raw state and in cooking is always gelatinized to some degree. For industrial purposes starches are submitted to many processes. Starch is often partially or almost wholly gelatinized or may be converted by heat or chemical treatment into dextrins for use in adhesive pastes, with the starch assuming a completely new form. Other treatments increase solubility, and hydrolysis with acids produces completely new products, including a variety of sugars.

Starch may be converted into sugars by the use of acids, and the sugars may be marketed as starch syrup, glucose syrup, or corn syrup; as glucose; and as commercial dextrose. Such sugars are useful in confectionery production.

Other uses of starch include production of ethyl alcohol by fermentation procedures and production of acetone and other products. Indeed, it is impossible to record all the hundreds of uses of starch in the science-based industries.

Alimentary Pastes

Pasta Products

Alimentary pastes include such products as macaroni, spaghetti, vermicelli, and noodles. Such products are often called pastas. Italy is regarded as the place of origin of macaroni products, and annual consumption in that country is as high as 30–35 kilograms (65–75 pounds) per person. Annual consumption is about 6.3 kilograms in France, 3.7 in the United States, and only 0.4 kilogram in the United Kingdom. Pasta is manufactured in a wide variety of sizes and shapes, the commonest being long, narrow strands. The most slender type of strand, vermicelli, sometimes called capelli d'angeli ("angel's hair") in Italy, has a diameter ranging from 0.5 to 0.8 millimetre and is normally cut into lengths of about 250 millimetres and twisted into curls. Short-cut vermicelli (15–40 millimetres) is easy to manufacture and to dry. Spaghetti has a diameter of about 1.5–2.5 millimetres and is usually straight. Noodles are solid ribbons, about 0.8 millimetre thick, and in a variety of widths. Macaroni is the commonest type of alimentary paste; it is hollow and has a greater thickness than the others. It can be shaped in a variety of forms, such as long, short, large, small tubes, etc.

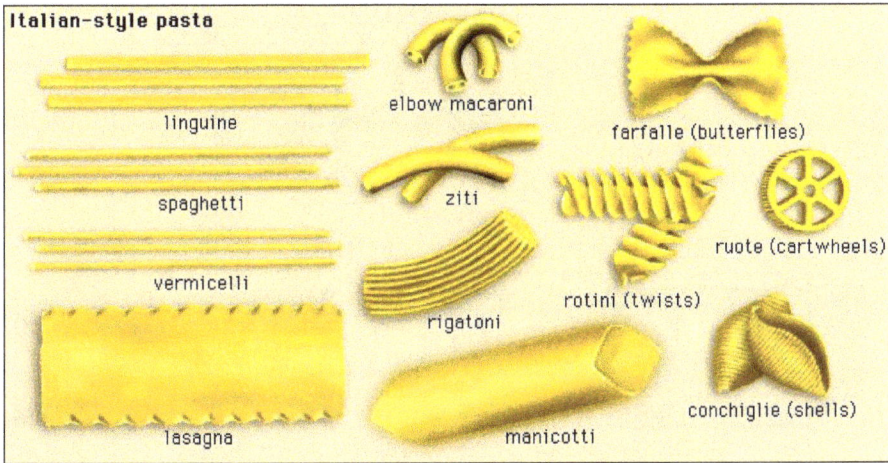

Italian-style pasta: linguine, spaghetti, vermicelli, lasagna, elbow macaroni, ziti, rigatoni, manicotti, farfalle (butterflies), ruote (cartwheels), rotini (twists), conchiglie (shells)

Macaroni is now commercially produced in large factories in Italy, North and South America, and other regions. Drying of the extruded paste is an important process, previously accomplished in Italy by sun drying.

Semolina

Semolina, not flour, is the form of cereal used, and various plain macaroni products are made by combining the correct form of semolina, from durum wheat, with water. Richer alimentary pastes are made with the addition of eggs in fresh, dried, or frozen form, and egg noodles are popular. In low-income families, alimentary pastes often provide the bulk of the calories in the diet. Macaroni products supply about 3,500 calories per kilogram and, although not themselves good sources of vitamins, are commonly cooked and consumed with butter, oil, cheese, and other items containing the needed vitamins.

The use of hard durum semolina contributes to good quality in macaroni and other alimentary paste products. The special mills involved use many breaks, and only a few reduction rolls, to produce as much clean semolina as possible. An efficient mill employing appropriate purifiers can produce as much as 65 percent semolina (together with a little flour). Before continuous processes for pasta production were introduced, a coarse semolina was valued. In modern production, semolina is dusted and freed from flour, and regularity in size is considered important for water absorption. Very fine semolina is not popular, and the preferred semolina usually has a moisture content of about 13 percent with less than 0.8 percent ash. Freedom from bran is desired to avoid the appearance of specks. The gluten in the semolina should be reasonably strong but not as elastic as that required for bread making.

Pasta Processing

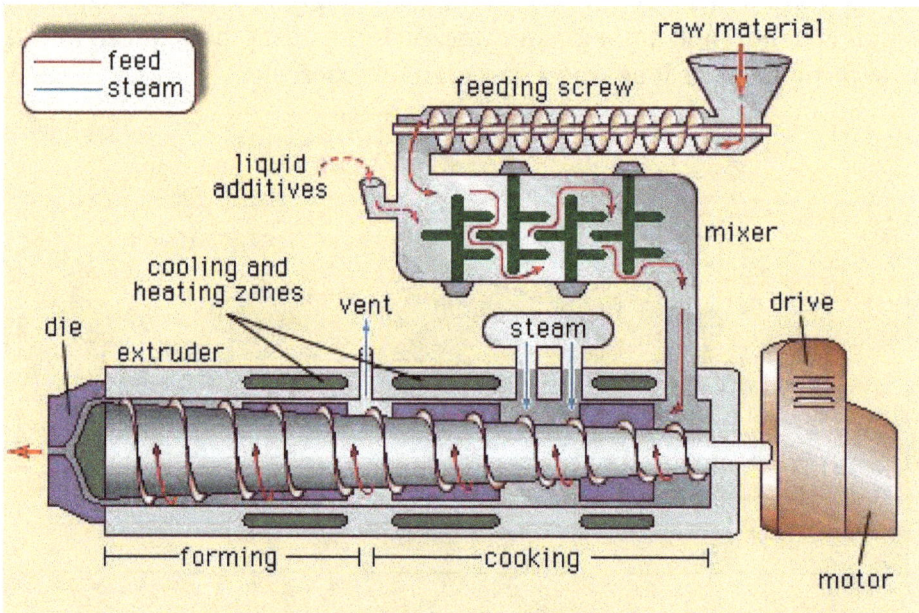

A high-temperature short-time extruder.

In the early factories, batch mixing of semolina and water was followed by extrusion of the resulting paste through presses containing dies. In modern practice, the bulk of alimentary pastes is made by continuous processes.

The basic procedure for most macaroni products consists of adding water to a semolina made from suitable wheat to produce, in a short time, a plastic homogeneous mass of about 30 percent moisture. This mixture is extruded through special dies, under pressure, producing the desired size and shape, and is then dried. There are many types of continuous paste processes adapted to the specific types of paste wanted and to the manufacturer's requirements. In the earlier days of the cottage industry, long-cut products such as spaghetti were spread evenly by hand on wooden dowels about an inch thick and over 50 inches long, and the filled sticks were then placed on racks for sun drying. Short-cut products were often scattered on wire mesh trays.

In modern automatic processing the objective is to dry the extruded product, containing 31 percent moisture, to a hard product of about 12 percent moisture, decreasing the possibility of the goods being affected by the growth of molds and yeast. If moisture is removed too rapidly, the dried product may tend to "check" or split. If moisture is removed too slowly, souring or mold growth may occur. Proper drying is therefore ensured by adjusting air circulation, temperature, and humidity. Drying procedures differ for long and short macaroni. In the continuous process, after a first hour in which a crust is formed to protect against mold infection, slow drying is practiced.

Testing

Cooking tests are used to ensure that the final product is satisfactory. Considerable research has been carried out to control factors tending to destroy the desirable yellow colour. Destruction of the colouring matter, a xanthophyll, can occur in mixing owing to excessive lipoxidase. Certain types of durum wheat may possess a high degree of lipoxidase activity, and it is difficult to control or check this action. The addition of ascorbic acid has been suggested as a means to decrease the destruction of the semolina pigments in processing.

In the United States, alimentary paste goods, described as noodles, egg spaghetti, or egg macaroni, must contain 5.5 percent of the solids of egg in the final product. The eggs can be used in the form of frozen yolks, dried yolks, frozen whole eggs, dried whole eggs, or fresh whole eggs or yolks. Spray-dried egg yolks of good quality are now available.

Breakfast Cereals

Origins

The modern packaged breakfast-food industry owes its beginnings to an American religious sect, the Seventh-day Adventists, who wished to avoid consumption of animal foods. In the 1860s they organized the Western Health Reform Institute in Battle Creek, Mich., later renamed the Battle Creek Sanitarium. James Jackson of Dansville, N.Y., produced a cereal food by baking whole-meal dough in thin sheets, breaking and regrinding into small chunks, rebaking and regrinding. J.H. Kellogg of Battle Creek made biscuits about one-half inch thick from a dough mixture of wheatmeal, oatmeal, and cornmeal. The dough was baked until it was fairly dry and turning brown, and the product was ground and packed. A patient at the sanitarium, C.W. Post, saw the possibilities in such a product entirely apart from the original conception of healthfulness and started a business. Kellogg's brother, W.K. Kellogg, did likewise, and the breakfast-food industry was launched, soon achieving mass sales of cereal products in flaked, granular, shredded, and puffed forms, with flavour obtained by roasting and the addition of sugar.

Types of Breakfast Cereal

Some breakfast cereals require cooking; others are packaged ready-to-eat. Roasted and rolled oatmeal, eaten as porridge, requires brief boiling. Cooking time of these processed cereals has been greatly reduced, and various "instant" forms are available.

Although cooked oatmeal porridge was formerly a standard breakfast food, the ready-to-eat cereals of various types are now the favourite breakfast-cereal foods.

The middlings produced in flour milling, essentially small pieces of endosperm free from bran and germ, are sold as farina and often consumed as a breakfast food in the United States. Farina is usually enriched with vitamins and minerals and may be flavoured. To reduce cooking time, 0.25 percent disodium phosphate may be added; some products require only one minute of boiling before serving.

Ready-to-eat cereals are available in a variety of forms and are normally consumed with milk and sometimes sugar. Flaked and toasted varieties are the most popular. During processing the starch is gelatinized, halting enzymatic reactions and thus ensuring product stability and good shelf life. The sugar content is dextrinized and caramelized by a roasting process. Roasting also ensures attractive crispness resulting from moisture reduction.

Flaked Cereals

Wheat and rice flakes are manufactured, but most flaked breakfast foods are made from corn (maize), usually of the yellow type, broken down into grits and cooked under pressure with flavouring syrup consisting of sugar, nondiastatic malt, and other ingredients. Cooking is often accomplished in slowly rotating retorts under steam pressure.

After leaving the cooker, the lumps (containing about 33 percent water) are broken down by revolving reels and sent to driers. These are usually large tubes extending vertically, through several stories, with the wet product entering the top and encountering a current of hot air (65 °C, or 150 °F). Other types of driers consist of horizontal rotating cylinders with steam-heated pipes running horizontally. The drying process reduces moisture to about 20 percent, and the product is transferred to tempering bins for up to 24 hours, to even moisture distribution.

The product is next flaked by passing it between large steel cylinders (180–200 revolutions per minute), with the rolls cooled by internal water circulation. The cooked and rather soft flakes then proceed to rotating toasting ovens (normally gas-fired), where the flakes tumble through perforated drums. This treatment requires two to three minutes at 225 °C (550 °F). The product is dehydrated, toasted, and slightly blistered. After toasting it is cooled by circulating air, and at this stage enrichment by sprays may be carried out.

The manufacture of wheat flakes is similar to that of corn flakes. Special machinery separates the individual grains so that they can be flaked and finally toasted.

Shredded Cereals

Shredded wheat, differing from other breakfast foods, is made from whole grains with the germ and bran retained and no flavour added. In its final form it is in tablets composed of shreds of cooked and toasted wheat. The wheat is cleaned and then boiled in water, often at atmospheric pressure. The grains reach a moisture content of 55 to 60 percent and require preliminary drying to about 50 percent. They are then placed in bins to condition them. The shredding process consists of passing the cooked and partially dried wheat to the shredding rolls, which are 150 to 200 millimetres (6 to 8 inches) in diameter and as wide as the finished tablet. On one pair of the rolls is a series of about 20 shallow corrugations running around the periphery; the surface of the other roll is smooth. The soft wheat is forced into the rolls under pressure and is cut into long shreds falling to a conveyor in such a way as to obtain superimposed shreds. These layers are cut into tablets by knives, and the tablets are transferred to baking pans. The pans pass to a revolving oven, with a baking temperature of approximately 260 °C (500 °F). After 10–15 minutes the outside of the product is dry and toasted, while the interior is still damp. The tablet is transferred either to another hot air oven or to a different section of the same oven, where it is dried at 120 °C (250 °F) for an additional 30 minutes and then cooled and packed.

Granular Cereals

Granular types are made by very different processes from the others. The first step is production of a stiff dough from wheat, malted barley flour, salt, dry yeast, and water. After mixing, fermentation proceeds for about five hours. The dough is then formed into large loaves and transferred directly to the oven. Baking requires about two hours at 205 °C (400 °F). The baked loaves are fragmented and the product is then thoroughly dried. Grinding by corrugated rolls follows, and the product is sieved to standard size. Very finely ground pieces are added to subsequent dough batches.

Puffed Cereals

Early in the 20th century an American patent was taken out for the preparation of puffed wheat and rice. Puffed oats and corn are now also produced. The principle of the puffing process is heating the cereal, and sometimes other vegetable products, in a pressure chamber to a pressure of 7 to 14 kilograms per square centimetre (100 to 200 pounds per square inch), then instantaneously releasing this pressure by suddenly opening the chamber, or puffing gun. Expansion of the water vapour occurs when the pressure is suddenly released, blowing up the grains or cereal pellets to several times their original size (8-fold to 16-fold for wheat, 6-fold to 8-fold for rice). The final product is toasted to a moisture content of about 3 percent to achieve desired crispness. In processing wheat, a preliminary step may be applied to free the grain from much of its bran coatings.

Rice is usually parboiled, pearled, and cooked with sugar syrup, dried to about 25 to 30 percent moisture in rotating louvre dryers, binned, and toasted and puffed. In puffing of mixed cereal products it is necessary to start with a stiff dough containing sugar, salt, and sometimes oil, and this mixture is then cooked. The dough is pelleted by extrusion through dies and dried to attain a suitable condition for the final puffing process.

Enrichment

Enrichment of breakfast cereals with minerals, and especially with vitamins, is now common practice. In many of the manufacturing processes employed in breakfast-food production, considerable vitamin destruction occurs. The various heat treatments involved may destroy 90 percent of the original B1 content of the cereal, especially in flaked and puffed products. On the other hand, a proportion of the somewhat harmful phytic acid in cereals, interfering with absorption by the body of calcium, is also destroyed; and enrichment of the products with vitamin B1, and sometimes other components of the vitamin B complex, is not difficult to perform after the various cooking operations have been completed.

Sweeteners

Various types of sweeteners are made directly from starch. Glucose products made by starch conversion differ in composition and in sweetness according to whether conversion is effected by acid or by enzymes. Enzyme-produced glucose is higher in dextrose and maltose content than acid-produced glucose, which normally is higher in dextrin. Sucrose is a more powerful sweetening agent than dextrose, but glucose syrup made by enzymatic treatment usually has twice the sweetening power of that produced by acid action.

In the production of starch separated by the wet milling of corn, one stream is normally used to produce starch, and the other stream is converted into corn syrup by heating the starch slurry in pressure tanks with acid or enzymes and following with refining processes. If the process of hydrolyzing starch is completed, the resulting product is glucose. Often the treatment is not carried to completion, and a series of dextrins and reversion products is produced. If full conversion is required, the treatment usually employs acids to liquefy, followed by saccharifying enzymes to complete the change to dextrose. Modern syrups and crystalline dextrose are made by continuous processes. The degree of conversion of the starch into the sugar dextrose is expressed as D.E. (dextrose equivalents), and confectionery syrups have a D.E. of about 36 to 55, while the fuller conversion of products with D.E. of 96 to 99 can be made for the production of almost pure glucose or dextrose, used in many food products.

Sweeteners in the form of syrups are largely used in cake and confectionery products and also, especially in the United States, in bread manufacture. American bread is distinctly sweeter than normal European bread because of the fats and sweeteners used, and the loaves are larger per unit of weight than in the United Kingdom and most European countries.

In the United States the baking industry uses more than one-half of the dextrose and 10 percent of the corn syrup produced. Makers of cookies (biscuits) and breakfast foods also use large amounts of sweeteners. Confectioners in Europe use syrups of many types but not as widely as in the United States.

MEAT PROCESSING

Meat processing is the preparation of meat for human consumption.

Meat is the common term used to describe the edible portion of animal tissues and any processed or manufactured products prepared from these tissues. Meats are often classified by the type of animal from which they are taken. Red meat refers to the meat taken from mammals, white meat refers to the meat taken from fowl, seafood refers to the meat taken from fish and shellfish, and game refers to meat taken from animals that are not commonly domesticated. In addition, most commonly consumed meats are specifically identified by the live animal from which they come. Beef refers to the meat from cattle, veal from calves, pork from hogs, lamb from young sheep, and mutton from sheep older than two years.

Conversion of Muscle to Meat

Muscle is the predominant component of most meat and meat products. Additional components include the connective tissue, fat (adipose tissue), nerves, and blood vessels that surround and are embedded within the muscles. The structural and biochemical properties of muscle are therefore critical factors that influence both the way animals are handled before, during, and after the slaughtering process and the quality of meat produced by the process.

Muscle Structure and Function

There are three distinct types of muscle in animals: smooth, cardiac, and skeletal. Smooth muscles, found in the organ systems including the digestive and reproductive tracts, are often used as casings for sausages. Cardiac muscles are located in the heart and are also often consumed as meat products. However, most meat and meat products are derived from skeletal muscles, which are usually attached to bones and, in the living animal, facilitate movement and support the weight of the body.

Skeletal Muscle Structure

Skeletal muscles are divided from one another by a covering of connective tissue called the epimysium. Individual muscles are divided into separate parts (called muscle bundles) by another connective tissue sheath known as the perimysium. Clusters of fat cells, small blood vessels (capillaries), and nerve branches are found in the region between muscle bundles. Muscle bundles are further divided into smaller cylindrical muscle fibres (cells) of varying lengths that are individually wrapped with a thin connective tissue sheath called the endomysium. Each of the connective tissue sheaths found throughout skeletal muscle is composed of collagen, a structural protein that provides strength and support to the muscles.

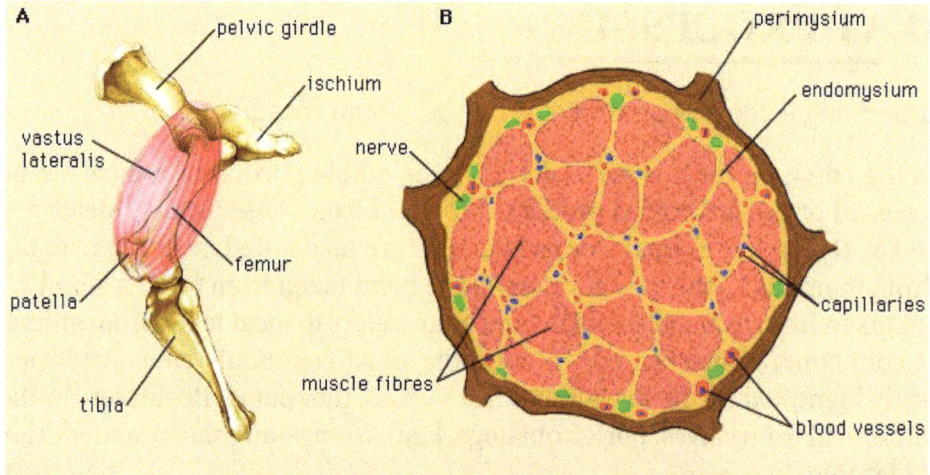

Drawing of the vastus lateralis muscle of sheep.

The plasma membrane of a muscle cell, called the sarcolemma, separates the sarcoplasm (muscle cell cytoplasm) from the extracellular surroundings. Within the sarcoplasm of each individual muscle fibre are approximately 1,000 to 2,000 myofibrils. Composed of the contractile proteins actin and myosin, the myofibrils represent the smallest units of contraction in living muscle.

Skeletal Muscle Contraction

The contraction of skeletal muscles is an energy-requiring process. In order to perform the mechanical work of contraction, actin and myosin utilize the chemical energy of the molecule adenosine triphosphate (ATP). ATP is synthesized in muscle cells from the storage polysaccharide glycogen, a complex carbohydrate composed of hundreds of covalently linked molecules of glucose (a monosaccharide or simple carbohydrate). In a working muscle, glucose is released from the glycogen reserves and enters a metabolic pathway called glycolysis, a process in which glucose is broken down and the energy contained in its chemical bonds is harnessed for the synthesis of ATP. The net production of ATP depends on the level of oxygen reaching the muscle. In the absence of oxygen (anaerobic conditions), the products of glycolysis are converted to lactic acid, and relatively little ATP is produced. In the presence of oxygen (aerobic conditions), the products of glycolysis enter a second pathway, the citric acid cycle, and a large amount of ATP is synthesized by a process called oxidative phosphorylation.

In addition to carbohydrates, fats supply a significant amount of energy for working muscles. Fats are stored in the body as triglycerides (also called triacylglycerols). A triglyceride is composed of three fatty acid molecules (nonpolar hydrocarbon chains with a polar carboxyl group at one end) bound to a single glycerol molecule. If the fat deposits are required for energy production, fatty acids are released from the triglyceride molecules in a process called fatty acid mobilization. The fatty acids are broken down into smaller molecules that can enter the citric acid cycle for the synthesis of ATP by oxidative phosphorylation. Therefore, the utilization of fats for energy requires the presence of oxygen.

An important protein of muscle cells is the oxygen-binding protein myoglobin. Myoglobin takes up oxygen from the blood (transported by the related oxygen-binding protein hemoglobin) and stores it in the muscle cells for oxidative metabolism. The structure of myoglobin includes a nonprotein group called the heme ring. The heme ring consists of a porphyrin molecule bound to an iron (Fe) atom. The iron atom is responsible for the binding of oxygen to myoglobin and has two possible oxidation states: the reduced, ferrous form (Fe^{2+}) and the oxidized, ferric form (Fe^{3+}). In the Fe^{2+} state iron is able to bind oxygen (and other molecules). However, oxidation of the iron atom to the Fe^{3+} state prevents oxygen binding.

Postmortem Muscle

Once the life of an animal ends, the life-sustaining processes slowly cease, causing significant changes in the postmortem (after death) muscle. These changes represent the conversion of muscle to meat.

pH Changes

Normally, after death, muscle becomes more acidic (pH decreases). When an animal is bled after slaughter (a process known as exsanguination), oxygen is no longer available to the muscle cells, and anaerobic glycolysis becomes the only means of energy production available. As a result, glycogen stores are completely converted to lactic acid, which then begins to build up, causing the pH to drop. Typically, the pH declines from a physiological pH of approximately 7.2 in living muscle to a postmortem pH of approximately 5.5 in meat (called the ultimate pH).

Protein Changes

When the energy reserves are depleted, the myofibrillar proteins, actin and myosin, lose their extendability, and the muscles become stiff. This condition is commonly referred to as rigor mortis. The time an animal requires to enter rigor mortis is highly dependent on the species (for instance, cattle and sheep take longer than hogs), the chilling rate of the carcass from normal body temperature (the process is slower at lower temperatures), and the amount of stress the animal experiences before slaughter.

Eventually the stiffness in the muscle tissues begins to decrease owing to the enzymatic breakdown of structural proteins (i.e., collagen) that hold muscle fibres together. This phenomenon is known as resolution of rigor and can continue for weeks after slaughter in a process referred to as aging of meat. This aging effect produces meats that are more tender and palatable.

Properties of Meat

Chemistry and Nutrient Composition

Regardless of the animal, lean muscle usually consists of approximately 21 percent protein, 73 percent water, 5 percent fat, and 1 percent ash (the mineral component of muscle). These figures vary as an animal is fed and fattened. Generally, as fat increases, the percentages of protein and water decrease. The table provides a comparison of the nutrient composition of many meat products.

Table: Nutrient composition of red meats (per 100 g).

Meat type and cut	Energy (kcal)	Water (g)	Protein (g)	Fat (g)	Cholesterol (mg)	Vitamin b_{12} (µg)	Thiamin (mg)	Iron (mg)	Zinc (mg)
Beef									
Chuck arm pot roast	219	58	33.02	8.70	101	3.40	0.080	3.79	8.66
Rib eye steak	225	59	28.04	11.70	80	3.32	0.100	2.57	6.99
Short ribs	295	50	30.76	18.13	93	3.46	0.065	3.36	7.80
Tenderloin	212	60	28.25	10.10	84	2.57	0.130	3.58	5.59
Top sirloin	200	61	30.37	7.80	89	2.85	0.130	3.36	6.52
Ground (extra lean)	265	54	28.58	15.80	99	2.56	0.070	2.77	6.43
Pork									
Loin roast	169	62	30.24	7.21	78	0.55	0.639	1.06	2.31
Tenderloin	164	66	8.14	4.81	79	0.55	0.940	1.47	2.63
Boston shoulder roast	232	61	24.21	14.30	85	0.93	0.669	1.56	4.23

Spareribs	397	40	29.06	30.30	121	1.08	0.382	1.85	4.60
Cured ham (extra lean)	145	68	20.93	5.53	53	0.65	0.754	1.48	2.88
Lamb								
Leg roast	191	64	28.30	7.74	89	2.64	0.110	2.12	4.94
Loin chop	202	63	26.59	9.76	87	2.16	0.100	2.44	4.06
Blade chop	209	63	24.61	11.57	87	2.74	0.090	2.07	6.48
Veal								
Loin chop	175	65	26.32	6.94	106	1.31	0.060	0.85	3.24
Rib chop	177	65	25.76	7.44	115	1.58	0.060	0.96	4.49

Protein

Meat is an excellent source of protein. As is explained above, these proteins carry out specific functions in living muscle tissue and in the conversion of muscle to meat. They include actin and myosin (myofibrillar proteins), glycolytic enzymes and myoglobin (sarcoplasmic proteins), and collagen (connective tissue proteins). Because the proteins found in meat provide all nine essential amino acids to the diet, meat is considered a complete source of protein.

Fat

Fats, in the form of triglycerides, accumulate in the fat cells found in and around the muscles of the animal. Fat deposits that surround the muscles are called adipose tissue, while fat that is deposited between the fibres of a muscle is called marbling.

In the diet the fats found in meat act as carriers for the fat-soluble vitamins (A, D, E, and K) and supply essential fatty acids (fatty acids not supplied by the body). In addition to their role as an energy reserve, fatty acids are precursors in the synthesis of phospholipids, the main structural molecules of all biological membranes.

Fatty acids are classified as being either saturated (lacking double bonds between their carbon atoms), monounsaturated (with one double bond), or polyunsaturated (containing several double bonds). The fatty acid composition of meats is dependent on several factors. In animals with simple stomachs, called nonruminants (e.g., pigs), diet can significantly alter the fatty acid composition of meat. If nonruminants are fed diets high in unsaturated fats, the fat they deposit in their muscles will have elevated levels of unsaturated fatty acids. In animals with multichambered stomachs, called ruminants (e.g., cattle and sheep), fatty acid composition found in the lean muscle is relatively unaffected by diet because microorganisms in the stomach alter the chemical composition of the fatty acids before they leave the digestive tract.

A beneficial characteristic of saturated fatty acids is that they do not undergo oxidation when exposed to air. However, the double bonds found in unsaturated fatty acids are susceptible to oxidation, and this oxidation promotes rancidity in meat. Therefore, products higher in saturated fats can generally be stored for a longer time without developing unpleasant flavours and odours.

Vitamins and Minerals

Meat contains a number of essential vitamins and minerals. It is an excellent source of many of the B vitamins, including thiamine, choline, B_6, niacin, and folic acid. Some types of meat, especially liver, also contain vitamins A, D, E, and K.

Meat is an excellent source of the minerals iron, zinc, and phosphorus. It also contains a number of essential trace minerals, including copper, molybdenum, nickel, selenium, chromium, and fluorine.

Cholesterol

Cholesterol is a constituent of cell membranes and is present in all animal tissues. Leaner meats typically are lower in cholesterol. Veal, however, is an exception: it is lower in fat than mature beef but has significantly higher cholesterol levels.

Carbohydrates

Meat contains virtually no carbohydrates. This is because the principal carbohydrate found in muscle, the complex sugar glycogen, is broken down in the conversion of muscle to meat.Liver is an exception, containing up to 8 percent carbohydrates.

Water

Water is the most abundant component of meat. However, because adipose tissue contains little or no moisture, as the percentage of fat increases in a meat cut, the percentage of water declines. Therefore, lean young veal may be as much as 80 percent water, while fully fattened beef may be as little as 50 percent. Because water is lost when meats are cooked, the percentages of protein and fat in cooked meats are usually higher than in their raw counterparts.

Colour

In well-bled animals approximately 80 to 90 percent of the total meat pigment is due to the oxygen-binding protein myoglobin. Colour differences in meat are related to the myoglobin content of muscle fibres and to the chemical state of the iron atom found in the myoglobin molecule.

Myoglobin Content

A number of factors influence the myoglobin content of skeletal muscles. Muscles are a mixture of two different types of muscle fibre, fast-twitch and slow-twitch, which vary in proportions between muscles. Fast-twitch fibres have a low myoglobin content and are therefore also called white fibres. They are dependent on anaerobic glycolysis for energy production. Slow-twitch fibres have a high amount of myoglobin and a greater capacity for oxidative metabolism. These fibres are often called red fibres. Therefore, dark meat colour is a result of a relatively high concentration of slow-twitch fibres in the muscle of the animal.

A second factor contributing to the myoglobin content of a muscle is the age of the animal—muscles from older animals often have higher myoglobin concentrations. This accounts for the darker colour of beef relative to that of veal.

The size of an animal may also affect the myoglobin content of its muscles because of differences in basal metabolic rates (larger animals have a lower metabolism). Some smaller animals (such as rabbits) typically have a lower myoglobin concentration (0.02 percent of wet weight of muscle) and lighter coloured meat than larger animals such as horses (0.7 percent myoglobin) or deep-diving animals such as whales, which have very high concentrations of myoglobin (7 percent myoglobin) and dark, purple-coloured meat. Myoglobin concentration is also greater in intact males (animals that have not been castrated) of similar age, in muscles located closer to the bones, and in more physically active animals such as game.

Oxidation State of Iron

The oxidation state of the iron atom of myoglobin also plays a significant role in meat colour. Meat such as beef viewed immediately after cutting is purple in colour because water is bound to the reduced iron atom of the myoglobin molecule (in this state the molecule is called deoxymyoglobin). Within 30 minutes after exposure to the air, beef slowly turns to a bright cherry-red colour in a process called blooming. Blooming is the result of oxygen binding to the iron atom (in this state the myoglobin molecule is called oxymyoglobin). After several days of exposure to air, the iron atom of myoglobin becomes oxidized and loses its ability to bind oxygen (the myoglobin molecule is now called metmyoglobin). In this oxidized condition, meat turns to a brown colour. Although the presence of this colour is not harmful, it is an indication that the meat is no longer fresh.

Tenderness

The tenderness of meat is influenced by a number of factors including the grain of the meat, the amount of connective tissue, and the amount of fat.

Meat Grain

Meat grain is determined by the physical size of muscle bundles. Finer-grained meats are more tender and have smaller bundles, while coarser-grained meats are tougher and have larger bundles. Meat grain varies between muscles in the same animal and between the same muscle in different animals. As a muscle is used more frequently by an animal, the number of myofibrils in each muscle fibre increases, resulting in a thicker muscle bundle and a stronger (tougher) protein network. Therefore, the muscles from older animals and muscles of locomotion (muscles used for physical work) tend to produce coarser-grained meat.

Connective Tissue

The amount of connective tissue in a muscle has a complex effect on the tenderness of the meat. The major component of connective tissue, collagen, has a tough, rigid structure. However, even though muscles from younger animals have more connective tissue, the meat derived from those muscles is generally more tender than that from older animals. This is due to the fact that collagen is broken down and denatured during the aging and cooking processes, forming a gelatin-like substance that makes the meat more tender. In addition, collagen becomes more rigid (resistant to breakdown and denaturation) with age, resulting in greater toughness of meat from older animals.

Fat

A high fat content within the adipose tissue and marbling sites of muscle contributes to the tenderness of the meat. During the cooking process the fat melts into a lubricant-type substance that spreads throughout the meat, increasing the tenderness of the final product.

Postmortem Quality Problems

Meat quality may be affected by both the preslaughter handling of the live animals and the postslaughter handling of the carcasses. Psychological or physical stress experienced by the animals produces biochemical changes in the muscles that may adversely affect the quality of the meat. In addition, postmortem muscles are susceptible to adverse biochemical reactions in response to certain external factors such as temperature.

DFD Meat

Dark, firm, and dry (DFD) meat is the result of an ultimate pH that is higher than normal. Carcasses that produce DFD meat are usually referred to as dark cutters. DFD meat is often the result of animals experiencing extreme stress or exercise of the muscles before slaughter. Stress and exercise use up the animal's glycogen reserves, and, therefore, postmortem lactic acid production through anaerobic glycolysis is diminished. The resulting postmortem pH of DFD meat is 6.2 to 6.5, compared with an ultimate pH value of 5.5 for normal meat. The dry appearance of this meat is thought to be a result of an unusually high water-holding capacity, causing the muscle fibres to swell with tightly held water. Because of its water content, this meat is actually juicier when cooked and eaten. Nevertheless, its dark colour and dry appearance result in a lack of consumer appeal, so that this meat is severely discounted at the marketplace.

PSE Meat

Pale, soft, and exudative (PSE) meat is the result of a rapid postmortem pH decline while the muscle temperature is too high. This combination of low pH and high temperature adversely affects muscle proteins, reducing their ability to hold water (the meat drips and is soft and mushy) and causing them to reflect light from the surface of the meat (the meat appears pale). PSE meat is especially problematic in the pork industry. It is known to be stress-related and inheritable. A genetic condition known as porcine stress syndrome (PSS) may increase the likelihood that a pig will yield PSE meat.

Cold Shortening

Cold shortening is the result of the rapid chilling of carcasses immediately after slaughter, before the glycogen in the muscle has been converted to lactic acid. With glycogen still present as an energy source, the cold temperature induces an irreversible contraction of the muscle (i.e., the actin and myosin filaments shorten). Cold shortening causes meat to be as much as five times tougher than normal. This condition occurs in lean beef and lamb carcasses that have higher proportions of red muscle fibres and very little exterior fat covering. Without the fat covering as insulation, the muscles can cool too rapidly before onset of rigor mortis. The process of electrical stimulation (the application of high-voltage electrical current to carcasses immediately postmortem) reduces or eliminates this condition by forcing muscle contractions and using up muscle glycogen. Thaw rigor is a similar condition that results when meat is frozen before it enters rigor mortis. When this meat is thawed, the leftover glycogen allows for muscle contraction and the meat becomes extremely tough.

Heat Ring

Heat ring is a problem associated with beef carcasses and results from differential chilling rates of the muscles after slaughter. A heat ring is a dark, coarsely textured band around the exterior portion of the muscle. In muscles that have a thin layer of external fat, the outer portion of the muscle may chill too fast after death, resulting in a slower pH decline in the outer layer and a dark-coloured ring. This condition is also alleviated by electrical stimulation of beef carcasses after slaughter, causing a more even pH decline throughout the muscle.

Livestock Slaughter Procedures

The slaughter of livestock involves three distinct stages: preslaughter handling, stunning, and slaughtering. In the United States the humane treatment of animals during each of these stages is required by the Humane Slaughter Act.

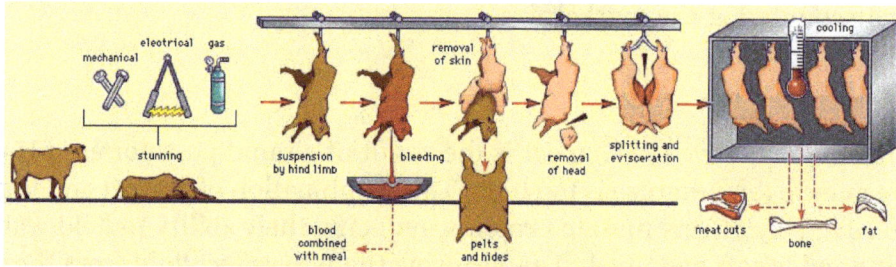

The basic slaughtering process.

Preslaughter Handling

Preslaughter handling is a major concern to the livestock industry, especially the pork industry. Stress applied to livestock before slaughter can lead to undesirable effects on the meat produced from these animals, including both PSE and DFD. Preslaughter stress can be reduced by preventing the mixing of different groups of animals, by keeping livestock cool with adequate ventilation, and by avoiding overcrowding. Before slaughter, animals should be allowed access to water but held off feed for 12 to 24 hours to assure complete bleeding and ease of evisceration (the removal of internal organs).

Stunning

As the slaughter process begins, livestock are restrained in a chute that limits physical movement of the animal. Once restrained, the animal is stunned to ensure a humane end with no pain. Stunning also results in decreased stress of the animal and superior meat quality.

The three most common methods of stunning are mechanical, electrical, and carbon dioxide (CO_2) gas. The end result of each method is to render the animal unconscious. Mechanical stunning involves firing a bolt through the skull of the animal using a pneumatic device or pistol. Electrical stunning passes a current of electricity through the brain of the animal. CO_2 stunning exposes the animal to a mixture of CO_2 gas, which acts as an anesthetic.

Slaughtering

After stunning, animals are usually suspended by a hind limb and moved down a conveyor line for the slaughter procedures. They are typically bled (a process called sticking or exsanguination) by the insertion of a knife into the thoracic cavity and severance of the carotid artery and jugular vein. This method allows for maximal blood removal from the body. At this point in the process, the slaughtering procedures begin to differ by species.

Hogs

Hogs are usually stunned by electrical means or CO_2 gas. Mechanical stunning is not generally used in hogs because it may cause serious quality problems in the meat, including blood splashing (small, visible hemorrhages in the muscle tissue) in the lean and PSE meat.

Hogs are one of the few domesticated livestock animals in which the skin is left on the carcass after the slaughter process. Therefore, after bleeding, the carcasses undergo an extensive cleaning procedure. First they are placed for about five minutes in a scalding tank of water that is between 57 and 63 °C (135 and 145 °F) in order to loosen hair and remove dirt and other material (called scurf) from the skin. The carcasses are then placed in a dehairing machine, which uses rubber paddles to remove the loosened hair. After dehairing, the carcasses are suspended from a rail with hooks placed through the gambrel tendons on the hind limbs, and any residual hair is shaved and singed off the skin.

An exception to this procedure occurs in certain specialized hog slaughter facilities, such as "whole hog" sausage slaughter plants. In whole hog sausage production all the skeletal meat is trimmed off the carcass, and therefore the carcass is routinely skinned following exsanguination.

After cleaning and dehairing, heads are removed and carcasses are opened by a straight cut in the centre of the belly to remove the viscera (the digestive system including liver, stomach, bladder, and intestines and the reproductive organs), pluck (thoracic contents including heart and lungs), kidneys, and associated fat (called leaf fat). The intestines are washed and cleaned to serve as natural casings for sausage products. The carcasses are then split down the centre of the backbone into two "sides," which are placed in a cooler (called a "hot box") for approximately 24 hours before fabrication into meat cuts.

Cattle, Calves and Sheep

These animals are usually stunned mechanically, but some sheep slaughter facilities also use electrical stunning. The feet are removed from the carcasses before they are suspended by the Achilles tendon of a hind leg for exsanguination. The carcasses are then skinned with the aid of mechanical skinners called "hide pullers." Sheep pelts are often removed by hand in a process called "fisting." (In older operations, hides and pelts are removed by knife.) The hides (cattle and calves) or pelts (sheep) are usually preserved by salting so that they can be tanned for leather products. Heads are removed at the first cervical vertebra, called the atlas joint. Evisceration and splitting are similar to hog procedures, except that kidney, pelvic, and heart fat are typically left in beef carcasses for grading. Carcasses are then placed in a cooler for 24 hours (often 48 hours for beef) prior to fabrication into meat cuts.

By-products

By-products are the nonmeat materials collected during the slaughter process, commonly called offal. Variety meats include livers, brains, hearts, sweetbreads (thymus and pancreas), fries (testicles), kidneys, oxtails, tripe (stomach of cattle), and tongue. Bones and rendered meat are used as bone and meat meal in animal feeds and fertilizers. Gelatin, obtained from high-collagen products such as pork snouts, pork skin, and dried rendered bone, is used in confections, jellies, and pharmaceuticals. Intestines are used as sausage casings. Hormones and other pharmaceutical products such as insulin, heparin, and cortisone are obtained from various glands and tissues. Edible fats are used as lard (from hogs), tallow (from cattle), shortenings, and cooking oils. Inedible fats are used in soap and candle manufacturing and in various industrial grease formulations. Lanolin from sheep wool is used in cosmetics. Finally, hides and pelts are used in the manufacture of leather.

Meat Inspection

Meat inspection is mandatory and has the mission of assuring wholesomeness, safety, and accurate labeling of the meat supply. Although inspection procedures vary from country to country, they are centred around the same basic principles and may be performed by government officials, veterinarians, or plant personnel. For example, in the United States meat inspection is administered through the Food Safety and Inspection Service of the United States Department of Agriculture (USDA-FSIS) and is composed of several distinct programs. In general, these programs are representative of the basic inspection procedures used throughout the world and include antemortem inspection, postmortem inspection, reinspection during processing, sanitation, facilities and equipment, labels and standards, compliance, pathology and epidemiology, residue monitoring and evaluation, federal-state relations, and foreign programs.

Antemortem and Postmortem Inspection

Antemortem inspection identifies animals not fit for human consumption. Here animals that are down, disabled, diseased, or dead (known as 4D animals) are removed from the food chain and labeled "condemned." Other animals showing signs of being sick are labeled "suspect" and are segregated from healthy animals for more thorough inspection during processing procedures.

Postmortem inspection of the head, viscera, and carcasses helps to identify whole carcasses, individual parts, or organs that are not wholesome or safe for human consumption.

Reinspection During Processing

Although previously inspected meat is used in the preparation of processed meat products, additional ingredients are added to processed meats. Reinspection during processing assures that only wholesome and safe ingredients are used in the manufacture of processed meat products (e.g., sausage and ham).

Sanitation

Sanitation is maintained at all meat-packing and processing facilities by mandatory inspection both before and during the production process. This includes floors, walls, ceilings, personnel, clothing, coolers, drains, equipment, and other items that come in contact with food products. In addition, all water used in the production process must be potable (reasonably free of contamination).

Facilities and Equipment

Facilities and equipment are inspected to ensure that they meet safety requirements. Facilities must have sufficient cooling and lighting, and rails from which carcasses are suspended must be high enough to assure that the carcasses never come in contact with the floor. Equipment must be able to be properly cleaned and must not adversely affect the wholesomeness of the products.

Labels and Standards

Labels and standards regulations assure that products are accurately labeled, that nutritional information meets requirements, and that special label claims (e.g., lean, light, natural) are accurate. Virtually all meat products must have the following components in their label: accurate product name, list of ingredients (in order of predominance), name and place of business of packer and manufacturer, net weight, inspection stamp and plant number, and handling instructions.

Compliance

Compliance assures that proper criminal, administrative, and civil sanctions are carried out against violators of food inspection laws. These violations include the sale of uninspected meat, the use of inaccurate labels, and the contamination of products.

Pathology and Epidemiology

Pathology and epidemiology programs support the efforts of meat inspectors by working with other public health agencies to minimize the risk from widespread food-poisoning outbreaks. These agencies work to identify the causative agents of food poisoning and prevent repeated occurrences by improving prevention techniques (e.g., proper handling and cooking and prevention of cross-contamination of raw and cooked products).

Residue Monitoring and Evaluation

Residue monitoring and evaluation programs identify animals containing harmful residues and remove them from the food chain. These residues include toxins from natural sources, from pesticides, from feeds, or from antibiotics administered to animals too soon before slaughter.

Meat Grading

Meat grading segregates meat into different classes based on expected eating quality (e.g., appearance, tenderness, juiciness, and flavour) and expected yield of salable meat from a carcass. In contrast to meat-inspection procedures, meat-grading systems vary significantly throughout the world. These differences are due in large part to the fact that different countries have different meat quality standards. For example, in the United States cattle are raised primarily for the production of steaks and are fattened with high-quality grain feed in order to achieve a high amount of marbling throughout the muscles of the animal. High marbling levels are associated with meat cuts that are juicier, have more flavour, and are more tender. Therefore, greater marbling levels—and especially marbling that is finely textured and evenly distributed—improve the USDA quality grade (i.e., Prime, Choice, or Select) of the beef. However, in Australia cattle are raised primarily for the production of ground beef products, and the highest quality grades are given to the leanest cuts of meat.

Some of the characteristics of meat used to assess quality and assign grades include: conformation of the carcass; thickness of external fat; colour, texture, and firmness of the lean meat; colour and shape of the bones; level of marbling; flank streaking; and degree of leanness.

Retail Meat Cutting

Meat vendor.

In the American style of meat cutting, whole carcasses are usually fabricated into more manageable primal (major) or subprimal (minor) cuts at the packing plant. This preliminary fabrication eases meat merchandising by reducing variability within the cuts. Primal and subprimal cuts are usually packaged and sold to retailers that further fabricate them into the products that are seen in the retail case.

Pork Fabrication

Hogs are slaughtered at approximately 108 kilograms (240 pounds) and yield carcasses weighing approximately 76 kilograms (70 percent yield of live weight). Pork carcasses are usually divided into two sides before chilling, and each side is divided into four lean cuts plus other wholesale cuts. The four lean cuts are the ham, loin, Boston butt (Boston shoulder), and picnic shoulder.

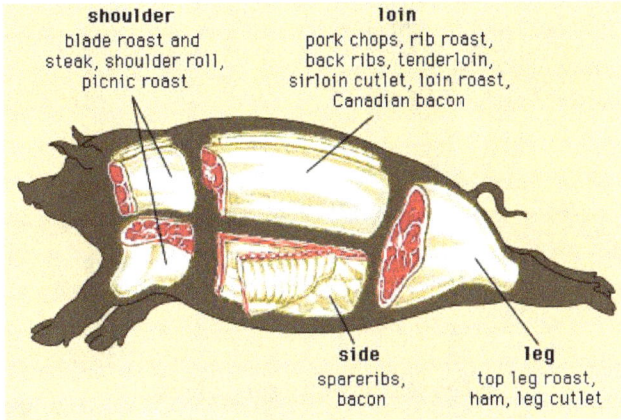

shoulder
blade roast and
steak, shoulder roll,
picnic roast

loin
pork chops, rib roast,
back ribs, tenderloin,
sirloin cutlet, loin roast,
Canadian bacon

side
spareribs,
bacon

leg
top leg roast,
ham, leg cutlet

Wholesale and retail cuts of pork.

Beef Fabrication

Steers and heifers average 495 kilograms at slaughter and produce carcasses weighing 315 kilograms (63 percent yield of live weight). Beef carcasses are split into two sides on the slaughter floor. After chilling, each side is divided into quarters, the forequarter and hindquarter, between the 12th and 13th ribs. The major wholesale cuts fabricated from the forequarter are the chuck, brisket, foreshank, rib, and shortplate. The hindquarter produces the short loin, sirloin, rump, round, and flank.

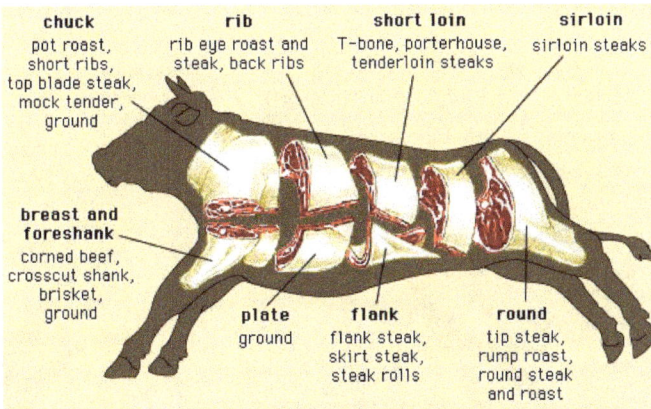

chuck
pot roast,
short ribs,
top blade steak,
mock tender,
ground

rib
rib eye roast and
steak, back ribs

short loin
T-bone, porterhouse,
tenderloin steaks

sirloin
sirloin steaks

**breast and
foreshank**
corned beef,
crosscut shank,
brisket,
ground

plate
ground

flank
flank steak,
skirt steak,
steak rolls

round
tip steak,
rump roast,
round steak
and roast

Wholesale and retail cuts of beef.

Lamb Fabrication

Live sheep averaging 45 kilograms yield 22-kilogram carcasses (50 percent yield of live weight). Lamb carcasses are divided into two halves, the foresaddle and hindsaddle, on the fabrication floor. The foresaddle produces the major wholesale cuts of the neck, shoulder, rib, breast, and foreshank. The hindsaddle produces the major wholesale cuts of the loin, sirloin, leg, and hindshank.

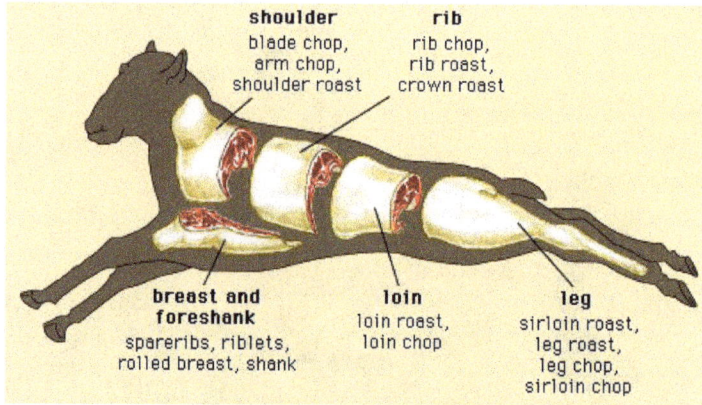

Wholesale and retail cuts of lamb.

Veal Fabrication

Veal is classified into several categories based on the ages of the animals at the time of slaughter. Baby veal (bob veal) is 2–3 days to 1 month of age and yields carcasses weighing 9 to 27 kilograms. Vealers are 4 to 12 weeks of age with carcasses weighing 36 to 68 kilograms. Calves are up to 20 weeks of age with carcasses ranging from 56 to 135 kilograms.

After slaughter, veal carcasses are split on the fabrication floor into two halves, the foresaddle and hindsaddle. The foresaddle produces the major wholesale cuts of the shoulder, rib, breast, and shank. The hindsaddle produces the major wholesale cuts of the loin, sirloin, and round.

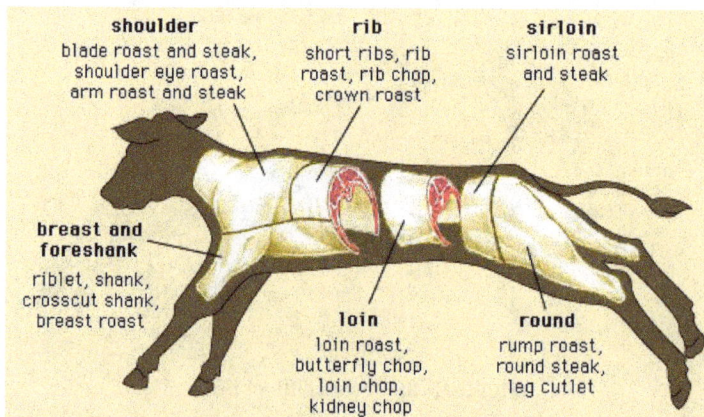

Wholesale and retail cuts of veal.

Meat Cookery

The physical changes associated with cooking meat are caused by the effects of heat on connective tissue and muscle proteins.

Colour Changes

In beef, changes in cooking temperatures ranging from 54 °C or 130 °F (very rare) to 82 °C or 180 °F correspond to changes in colour from deep red or purple to pale gray. These colour changes are a result of the denaturation of the myoglobin in meat. Denaturation is the physical unfolding of proteins in response to such influences as extreme heat. The denaturation of myoglobin makes the protein unable to bind oxygen, causing the colour to change from the bright cherry red of oxymyoglobin to the brown of denatured myoglobin (equivalent to metmyoglobin).

Structural Changes

The colour changes during cooking correspond to structural changes taking place in the meat. These structural changes are due to the effects of heat on collagen (connective tissue protein) and actin and myosin (myofibrillar proteins). In the temperature range between 50 and 71 °C (122 to 160 °F) connective tissue in the meat begins to shrink. Further heating to temperatures above 71 °C causes the complete denaturation of collagen into a gelatin-like consistency. Therefore, tough meats with relatively high amounts of connective tissues can be slowly cooked under moist conditions to internal temperatures above 71 °C and made tender by gelatinization of the collagen within the meat, while at the same time maintaining juiciness.

The myofibrillar proteins also experience major changes during cooking. In the range of 40 to 50 °C (104 to 122 °F) actin and myosin begin to lose solubility as heat denaturation begins. At temperatures of 66° to 77 °C (150 to 170 °F) the myofibrillar proteins begin to shorten and toughen. Beyond 77 °C (170 °F) proteins begin to lose structural integrity (i.e., they are completely denatured) and tenderness begins to improve.

The effects of heat on both connective tissue and myofibrillar proteins must be balanced in order to achieve maximum tenderness during cooking. Meats with low amounts of connective tissue are most tender when served closer to medium rare or rare so that muscle proteins are not hardened. Conversely, meats with heavy amounts of connective tissue require slow cooking closer to well done in order to achieve collagen gelatinization.

Meat Microbiology, Safety and Storage

When the conversion of muscle to meat begins, biological degradation of meat also commences. In the absence of a living immune system, microorganisms are unchecked in their ability to grow and reproduce on meat surfaces.

Food-borne Microorganisms

Generally, food-borne microorganisms can be classified as either food-spoilage or food-poisoning, with each presenting unique characteristics and challenges to meat product safety and quality.

Food-spoilage Microorganisms

These organisms are responsible for detrimental quality changes in meat. The changes include discoloration, unpleasant odours, and physical alterations. The principal spoilage organisms are molds and bacteria.

Molds usually appear dry and fuzzy and are white or green in colour. They can impart a musty flavour to meat. Common molds in meat include the genera Cladosporium, Mucor, and Alternaria. Slime molds produce a soft, creamy material on the surface of meat.

Common spoilage bacteria include Pseudomonas, Acinetobacter, and Moraxella. Under anaerobic conditions, such as in canned meats, spoilage can include souring, putrefaction, and gas production. This is a result of anaerobic decomposition of proteins by the bacteria.

Food-poisoning Microorganisms

Food-poisoning microorganisms can cause health problems by either intoxication or infection. Intoxication occurs when food-poisoning microorganisms produce a toxin that triggers sickness when ingested. Several different kinds of toxins are produced by the various microorganisms. These toxins usually affect the cells lining the intestinal wall, causing vomiting and diarrhea. Microorganisms capable of causing food-poisoning intoxication include Clostridium perfringens (found in temperature-abused cooked meats—i.e., meats that have not been stored, cooked, or reheated at the appropriate temperatures), Staphylococcus aureus (found in cured meats), and Clostridium botulinum (found in canned meats).

Infection occurs when an organism is ingested by the host, then grows inside the host and causes acute sickness and, in extreme cases, death. Common infectious bacteria capable of causing food poisoning in undercooked or contaminated meats are Salmonella, Escherichia coli, Campylobacter jejuni, and Listeria monocytogenes.

Prevention of Microbial Contamination

The initial microorganism load can be the most significant factor affecting the contamination of meat. If meat is never exposed to pathogenic microorganisms (those capable of causing human sickness), then there is no opportunity for food-borne illnesses to occur.

Several meat-processing plants have begun to utilize a program called the Hazard Analysis and Critical Control Point (HACCP) system to reduce pathogenic contamination. This program identifies the steps in the conversion of livestock to human food where the product is at risk of contamination by microorganisms. Once identified, these points, known as critical control points, are examined to determine how to eliminate the risk of microbial contamination.

Preservation and Storage

Meat preservation helps to control spoilage by inhibiting the growth of microorganisms, slowing enzymatic activity, and preventing the oxidation of fatty acids that promote rancidity. There are many factors affecting the length of time meat products can be stored while maintaining product safety and quality. The physical state of meat plays a role in the number of microorganisms that can grow on meat. For example, grinding meat increases the surface area, releases moisture and nutrients from the muscle fibres, and distributes surface microorganisms throughout the meat. Chemical properties of meat, such as pH and moisture content, affect the ability of microorganisms to grow on meat. Natural protective tissues (fat or skin) can prevent microbial contamination, dehydration, or other detrimental changes. Covering meats with paper or protective plastic films prevents excessive moisture loss and microbial contamination.

Cold Storage

Temperature is the most important factor influencing bacterial growth. Pathogenic bacteria do not grow well in temperatures under 3 °C (38 °F). Therefore, meat should be stored at temperatures that are as cold as possible. Refrigerated storage is the most common method of meat preservation. The typical refrigerated storage life for fresh meats is 5 to 7 days.

Freezer storage is an excellent method of meat preservation. It is important to wrap frozen meats closely in packaging that limits air contact with the meat in order to prevent moisture loss during storage. The length of time meats are held at frozen storage also determines product quality. Under typical freezer storage of −18 °C (0 °F) beef can be stored for 6 to 12 months, lamb for 6 to 9 months, pork for 6 months, and sausage products for 2 months.

Freezing

The rate of freezing is very important in maintaining meat quality. Rapid freezing is superior; if meats are frozen slowly, large ice crystals form in the meat and rupture cell membranes. When this meat is thawed, much of the original moisture found in the meat is lost as purge (juices that flow from the meat). For this reason cryogenic freezing (the use of supercold substances such as liquid nitrogen) or other rapid methods of freezing meats are used at the commercial level to maintain maximal product quality. It is important to note, however, that freezing does not kill most microorganisms; they simply become dormant. When the meat is thawed, the spoilage continues where it left off.

Thawing meats often can cause more detrimental quality changes than freezing. In contrast to freezing, thawing should be a slow process. Meats are best thawed in the refrigerator with packaging left intact, so that moisture loss is minimized. Placing frozen meats out on a warm countertop or under warm water subjects the meat's outer layers to room temperatures for long periods of time before the meat is ready for cooking (completely thawed). This rapid method provides a conducive environment for the growth of food-borne microorganisms and increases the risk of food poisoning.

Vacuum Packaging

Oxygen is required for many bacteria to grow. For this reason most meats are vacuum-packaged, which extends the storage life under refrigerated conditions to approximately 100 days. In addition, vacuum packaging minimizes the oxidation of unsaturated fatty acids and slows the development of rancid meat.

Canning

The second most common method of meat preservation is canning. Canning involves sealing meat in a container and then heating it to destroy all microorganisms capable of food spoilage. Under normal conditions canned products can safely be stored at room temperature indefinitely. However, certain quality concerns can compel processors or vendors to recommend an optimal "sell by" date.

Drying

Drying is another common method of meat preservation. Drying removes moisture from meat products so that microorganisms cannot grow. Dry sausages, freeze-dried meats, and jerky products are all examples of dried meats capable of being stored at room temperature without rapid spoilage.

Fermentation

One ancient form of food preservation used in the meat industry is fermentation. Fermentation involves the addition of certain harmless bacteria to meat. These fermenting bacteria produce acid as they grow, lowering the pH of the meat and inhibiting the growth of many pathogenic microorganisms.

Irradiation

Irradiation, or radurization, is a pasteurization method accomplished by exposing meat to doses of radiation. Radurization is as effective as heat pasteurization in killing food-spoilage microorganisms. Irradiation of meat is accomplished by exposing meat to high-energy ionizing radiation produced either by electron accelerators or by exposure to gamma-radiation-emitting substances such as cobalt-60 or cesium-137. Irradiated products are virtually identical in character to nonirradiated products, but they have significantly lower microbial contamination. Irradiated fresh meat products still require refrigeration and packaging to prevent spoilage, but the refrigerated storage life of these products is greatly extended.

Curing and Smoking

Meat curing and smoking are two of the oldest methods of meat preservation. They not only improve the safety and shelf life of meat products but also enhance the colour and flavour. Smoking of meat decreases the available moisture on the surface of meat products, preventing microbial growth and spoilage. Meat curing, as commonly performed in products such as ham or sausage, involves the addition of mixtures containing salt, nitrite, and other preservatives.

Salt decreases the moisture in meats available to spoilage microorganisms. Nitrite prevents microorganisms from growing and retards rancidity in meats. Nitrite also produces the pink colour associated with cured products by binding (as nitric oxide) to myoglobin. However, the use of nitrite in meat products is controversial owing to its potential cancer-causing activity.

Sodium erythorbate or ascorbate is another common curing additive. It not only decreases the risks associated with the use of nitrite but also improves cured meat colour development. Other common additives include alkaline phosphates, which improve the juiciness of meat products by increasing their water-holding ability.

PERMISSIONS

All chapters in this book are published with permission under the Creative Commons Attribution Share Alike License or equivalent. Every chapter published in this book has been scrutinized by our experts. Their significance has been extensively debated. The topics covered herein carry significant information for a comprehensive understanding. They may even be implemented as practical applications or may be referred to as a beginning point for further studies.

We would like to thank the editorial team for lending their expertise to make the book truly unique. They have played a crucial role in the development of this book. Without their invaluable contributions this book wouldn't have been possible. They have made vital efforts to compile up to date information on the varied aspects of this subject to make this book a valuable addition to the collection of many professionals and students.

This book was conceptualized with the vision of imparting up-to-date and integrated information in this field. To ensure the same, a matchless editorial board was set up. Every individual on the board went through rigorous rounds of assessment to prove their worth. After which they invested a large part of their time researching and compiling the most relevant data for our readers.

The editorial board has been involved in producing this book since its inception. They have spent rigorous hours researching and exploring the diverse topics which have resulted in the successful publishing of this book. They have passed on their knowledge of decades through this book. To expedite this challenging task, the publisher supported the team at every step. A small team of assistant editors was also appointed to further simplify the editing procedure and attain best results for the readers.

Apart from the editorial board, the designing team has also invested a significant amount of their time in understanding the subject and creating the most relevant covers. They scrutinized every image to scout for the most suitable representation of the subject and create an appropriate cover for the book.

The publishing team has been an ardent support to the editorial, designing and production team. Their endless efforts to recruit the best for this project, has resulted in the accomplishment of this book. They are a veteran in the field of academics and their pool of knowledge is as vast as their experience in printing. Their expertise and guidance has proved useful at every step. Their uncompromising quality standards have made this book an exceptional effort. Their encouragement from time to time has been an inspiration for everyone.

The publisher and the editorial board hope that this book will prove to be a valuable piece of knowledge for students, practitioners and scholars across the globe.

INDEX

www.ingramcontent.com/pod-product-compliance
Lightning Source LLC
Chambersburg PA
CBHW061948190326
41458CB00009B/2815